SpringerWienNewYork

Clinical Aspects and Laboratory

Electrolytes, Acid-Base Balance and Blood Gases

W.-R. Külpmann
H.-K. Stummvoll †
P. Lehmann

Second, enlarged and revised edition

SpringerWienNewYork

Prof. Dr. Wolf-Rüdiger Külpmann
Klinische Chemie, Medizinische Hochschule Hannover
Hannover, Germany

Prim. Univ.-Doz. Dr. Hans-Krister Stummvoll †
Interne Abteilung mit Nephrologie, Krankenhaus der Elisabethinen
Linz, Austria

Dr. Paul Lehmann
Roche Diagnostics GmbH
Mannheim, Germany

Enlarged and revised edition of Electrolytes, Clinical and Laboratory Aspects
Springer-Verlag/Wien 1996

This work is subject to copyright.
All rights are reserved, whether the whole or part of the material is concerned, specifically those of translation, reprinting, re-use of illustrations, broadcasting, reproduction by photocopying machines or similar means, and storage in data banks.

© 1996 and 2007 Springer-Verlag/Wien
Printed in Austria

SpringerWienNew York is part of
Springer Science + Business Media
springer.com

Product Liability: The publisher can give no guarantee for the information contained in this book. This also refers to that on drug dosage and application thereof. In each individual case the respective user must check the accuracy of the information given by consulting other pharmaceutical literature. The use of registered names, trademarks, etc. in this publication does not imply, even in the absence of a specific statement, that such names are exempt from the relevant protective laws and regulations and therefore free for general use.

Typesetting: Composition & Design Services, Minsk, Belarus
Printing and binding: Holzhausen Druck und Medien GmbH, 1140 Vienna, Austria

Printed on acid-free and chlorine-free bleached paper
SPIN: 11693680

With 45 Figures

CIP data applied for

ISBN-10   3-211-33127-1 SpringerWienNewYork
ISBN-13   978-3-211-33127-9 SpringerWienNewYork
ISBN 3-211-82790-0 1st ed. SpringerWienNewYork

Preface

This book examines the medical significance of electrolytes, acids, bases and blood gases, and how they are determined. It was written to provide physicians with a snapshot of the analytical testing of electrolytes, acids, bases and blood gases. In addition, it provides laboratory technicians with an overview of physiology and pathophysiology in these fields.

The first part of the book provides a summary of the current status of diagnosis and therapy. It can be used for quick reference at the patient's bedside or to more thoroughly research pathophysiological interdependencies. A separate section is dedicated to electrolytes in urine.

The latter chapters discuss the preanalytical and analytical testing of electrolytes, the acid-base balance and blood gases, with special consideration for performing determination using ion-selective electrodes and carrier-bound reagents ("dry chemistry"). The final chapter discusses the quality assurance of the methods, with consideration for the new Guidelines of the German Bundesärztekammer (2001), which are presented in outline form.

October 2006
W.-R. Külpmann
H.-K. Stummvoll †
P. Lehmann

Contents

Introduction 1

1. Electrolytes in the Serum 6
1.1 Physiological and Pathophysiological Principles of Electrolyte Balance 6
1.2 Sodium 8
1.3 Chloride 22
1.4 Anion Gap in the Serum 25
1.5 Osmolality 28
1.6 Potassium 30
1.7 Magnesium 38
1.8 Calcium 45
1.9 Phosphate 54

2. Electrolytes in the Urine 60
2.1 Physiological and Pathophysiological Principles 60
2.2 Sodium Excretion 64
2.3 Chloride Excretion 67
2.4 Osmolality 68
2.5 Potassium Excretion 70
2.6 Anion Gap 73
2.7 Magnesium Excretion 74
2.8 Calcium Excretion 76
2.9 Phosphate Excretion 79

3. Acid-Base Balance and Blood Gases 85
3.1 Physiology of Acid-Base Balance 85
3.2 Pathophysiology of Acid-Base Balance: Introduction 89
3.3 Metabolic Acidosis 92
3.4 Metabolic Alkaloses 96
3.5 Respiratory Acidoses 98
3.6 Respiratory Alkaloses 99

3.7	Treatment	100
3.8	Oxygen: Physiology	102
3.9	Oxygen: Pathophysiology	110

4. Preanalysis ... 114
4.1	Sodium	114
4.2	Chloride	114
4.3	Osmolality	115
4.4	Potassium	115
4.5	Magnesium	117
4.6	Total Calcium	117
4.7	Ionized Calcium	118
4.8	Ionized Phosphate	118
4.9	Electrolyte Determinations in Urine	119
4.10	Acid-Base Balance in Blood and Blood Gases	120
4.11	Acid-Base Measurement in Urine	123

5. Methods of Determination ... 124
5.1	Ion-Selective Electrodes	126
5.2	Absorption Spectrometry – Photometric Determination	140
5.3	Flame Atomic Emission Spectrometry ("Flame Photometry"; Sodium, Potassium, Calcium)	143
5.4	Atomic Absorption Spectrometry (Magnesium, Calcium)	147
5.5	Coulometry (Chloride)	149
5.6	Osmometry (Osmolality)	150
5.7	Carrier-Bound Reagents ("Dry-Phase Technology")	150

6. Methods of Determination of Acid-Bases and Blood Gases ... 153
6.1	Acid-Base Balance in Blood	153
6.2	Blood Gases	158
6.3	Acid-Base Balance in Urine	162

7. Quality Assurance ... 164

Appendix ... 174

References ... 180
  Review Literature ... 190

Elements ... 191

Introduction

As clinical symptoms are rather unspecific, diagnostic indications of disorders of water or electrolyte balance are provided primarily by the serum and urine electrolyte values for sodium, chloride, potassium, magnesium, calcium, phosphate and by determination of osmolality.

Sodium is quantitatively the most important extracellular cation, playing a decisive role in the fluid balance of the body. Chloride is also found almost exclusively in the extracellular compartment and, as the main anion, follows the sodium cations. In the majority of cases the changes in chloride concentration parallel those of the sodium concentration.

Serum osmolality is mainly dependent on low-molecular substances, especially sodium, chloride and bicarbonate ions as well as urea and glucose. In uremia, hyperglycemia, hyperlactatemia or alcohol intoxication the increase in osmotically active molecules leads to an increase in osmolality in the serum and thus to a shift of water from the intracellular compartment to the extracellular compartment.

The measurement of osmolality provides through calculation of the osmolal gap, amongst other things, indications of intoxication, e.g. with ethanol, in addition to information on the electrolyte balance.

An important regulator of water distribution between intra- and extracellular space is albumin. It is the substance mainly responsible for the maintenance of colloid osmotic pressure. As colloid osmotic pressure can be measured directly, the direction and extent of the shift in fluids can be recognized in a short time.

Disorders of potassium balance threaten the patient's life especially because of neuromuscular effects, in particular in the myocardium. Potassium, as the quantitatively most important cation of the intracellular compartment, determines the osmotic relationships within the cell. There is an immense concentration gradient between intracellular and extracellular space, so that even small percentage losses of potassium from the cell can lead to a marked increase in the concentration in extracellular space. The electric potential across the cell membrane, and thus muscular contractility and the conductivity of nerve cells, is influenced by the potassium concentration and is decisive for clinical symptoms in the event of potassium imbalance.

As the potassium and H^+ ions are exchangeable intracellulary, there is a close relationship between potassium concentration and acid-base balance. In acidosis (H^+-excess), potassium ions in exchange with H^+-ions flow from intra- to extracellular space and this leads to hyperkalemia. The conditions are the reverse in alkalosis. A normal potassium concentration in the serum therefore means potassium deficiency in severe acidosis or potassium excess in severe alkalosis.

Next to potassium, magnesium is the quantitatively most important intracellular cation. Magnesium activates over 300 enzymes. There are indications to suggest that latent magnesium deficiency is common in the general population. Manifest deficiency leads to gastrointestinal, neurological and cardiac symptoms.

Calcium ions influence the course of contraction of the heart and of skeletal muscle, and are essential for the functioning of the nervous system. In addition, they play an important role in blood clotting and are responsible for bone mineralization. In the plasma, calcium is bound to a considerable degree to protein (about 40%). This protein binding is pH-dependent. 10% is present in the form of inorganic complexes and 50% is free (known as "ionized" calcium). The citrate complex plays an important role in the incorporation and release of calcium from bones. A shift in the ratio of free to bound calcium occurs especially in neonates and premature children (acidosis, hypo-osmolality) or after massive transfusions (binding of calcium by the citrate present in the blood units). In such cases it is advisable to determine not only total calcium, but also ionized (free) calcium.

In critical care medicine, in particular, disturbances of acid-base balance and oxygen delivery play a very important role. Hydrogen ion activity, which is reflected in the pH, is tightly regulated. This regulation involves the various buffering systems in the blood (e.g. the bicarbonate and phosphate buffers and hemoglobin) as well as the lungs and the kidneys. A state in which the hydrogen ion activity is increased (pH reduced) is referred to as acidosis. Depending on the cause, this is further described as respiratory or metabolic acidosis. A state in which hydrogen ion activity is reduced (pH raised) is referred to as alkalosis. If effective counterregulation has taken place we speak of compensated acidosis or alkalosis.

The main factor influencing oxygen (O_2) delivery to the tissues is lung function. Alterations of lung function are referred to as disturbances of ventilation, perfusion, distribution or diffusion, depending on the predominant pathomechanism. Oxygen delivery is also impaired if the proportion of dyshemoglobins or abnormal hemoglobins is increased (e.g. persistence of hemoglobin F, hemoglobin Zürich, thalassemia).

Clinical symptoms of disorders of electrolyte balance are generally discrete, ambiguous and frequently only apparent when the patient's life is already seriously threatened. For these reasons, the monitoring of electrolyte concentrations form part of the basic examination in patient care.

The preferred methods of measurement for determining electrolytes are currently flame atomic emission spectrometry (FAES), flame atomic absorption spectrometry (FAAS) and potentiometry using ion-selective electrodes (ISE) as well as coulometry for chloride determination.

When examining samples by means of ion-selective electrodes, one must differentiate between measurements on the undiluted sample and measurements following dilution of the sample. The analysis of undiluted serum yields a measure for the ion activity in the native sample material.

With marked dilution of the serum, the ionic strength of the sample is adjusted to the ionic strength of the calibration solution and the electrolyte-free compartment of the macromolecules is reduced to less than 1% of the total volume. In this measurement technique, the measurement signals are duly converted by comparison with calibration solutions into concentrations. The concentration values obtained are identical to the results obtained by means of flame atomic emission spectrometry, if binding is negligible. This means that the accuracy of analyses by ion-selective electrodes after dilution of the sample can be checked by the reference method values obtainable for the total amount of electrolyte concentration in a sample.

The quantity related to "direct" ISE (i.e. measurement in undiluted samples) is the ion activity of the free ion of a given species in the water phase, which is independent of the amount of cells, proteins and lipids present in the solution. The regular monitoring of electrolytes is urgently required especially in patients who are being treated extensively with infusions as massive deviations in isoiony and isotonicity can occur with this therapy. The increasing need in this field for rapid determinations of the pertinent quantities has been taken into account by the introduction of "direct" ISEs, which allow analytical results to be obtained within a few minutes because the blood centrifugation required to obtain plasma, or additional waiting for clotting when using serum, is unnecessary through the use of heparinized blood. The electrode measurement takes place in the extracellular water phase of the blood or in the plasma or serum water and thus permits an interpretation of the results without knowledge of the protein and lipid content of the sample, which is of importance in the event of small relative reference intervals (e.g. sodium).

Most analyzers used in clinical chemistry contain an absorption spectrometer as a central module. The first thing that comes to mind therefore is to look for methods which permit the determination of electrolytes by absorption spectrometry. Developments in recent years have brought major advances in this direction. Methods in which electrolytes are determined "enzymatically" deserve special mention. In these procedures the electrolytes act as effectors of an enzyme-catalyzed reaction in which the pertinent substrate is more or less rapidly metabolized depending on the concentration of the effector.

Rapid potassium determinations in the blood have recently become possible with carrier-bound reagents ("dry chemistry").

The parameters of acid base balance and oxygen delivery measured in the blood are pH (negative logarithm of the relative molar hydrogen ion activity), pCO_2 (CO_2 partial pressure) and pO_2 (O_2 partial pressure). All other parameters, e.g. total CO_2 (plasma), actual bicarbonate (plasma) and base excess, are calculated using algorithms which are sometimes very complex. At present O_2 saturation (blood) is still usually calculated, but direct measurement of this parameter is increasing. The same applies to the O_2 hemoglobin fraction (blood) and total O_2 concentration (blood).

pH, pCO_2 and pO_2 are determined potentiometrically (pH, pCO_2) or amperometrically (pO_2) using special electrodes. Blood gas analyzers usually contain additional electrodes for simultaneous determination of sodium (ionized), potassium (ionized) and calcium (ionized) and frequently also for measurement of glucose, lactate and chloride (ionized).

Laboratories monitor and document the reliability of their methods by means of quality control and plausibility checks (examination of trends, consistency with other findings, extreme values). On the other hand, errors occurring in the preanalytical phase, e.g. during specimen collection or transport, are usually beyond the control of the laboratory. Yet these errors often lead to deviations which are far greater than those caused by imprecision, inaccuracy or unspecificity of the methods. Pre-analysis is therefore dealt with here in particular depth.

1. Electrolytes in the Serum

1.1 Physiological and Pathophysiological Principles of Electrolyte Balance

Water accounts for about 60% of the body weight (BW); two thirds of total water are attributed to the intracellular space (ICS) and one third is located in the extracellular space (ECS). The two spaces are separated from one another by the cell membrane. The extracellular space is subdivided into the interstitial space, ISS, and intravascular space, IVS. The ISS contains about 75% and the IVS about 25% of the water of the ECS. The border between these two spaces is the capillary wall (Fig. 1).

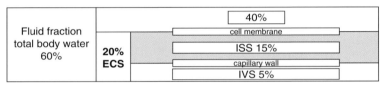

Fig. 1. Fluid spaces and fluid distribution in the body compartments (data as % of body weight). ECS: Extracellular space. ICS: Intracellular space. ISS: Interstitial space. IVS: Intravascular space.

It is difficult to investigate the intracellular space directly. It is, however, undoubtedly nowhere as uniform as the term "ICS" suggests. Every type of cell has its own "internal life" which it regulates itself.

The intracellular space contains a high potassium concentration, and phosphate is the predominant anion (Fig. 2). As the quantitatively

Extracellular Space				Intracellular Space			
Na$^+$	140	Cl$^-$	100	K$^+$	160	HPO$_4^{2-}$	50
K$^+$	4	HPO$_4^{2-}$	1	Na$^+$	10	Cl$^-$	3
Ca^{2+}	2.5	HCO$_3^-$	27	Ca^{2+}	1	HCO$_3^-$	10
Mg^{2+}	1	SO$_4^{2-}$	0.5	Mg^{2+}	13	SO$_4^{2-}$	10

Fig. 2. Concentration of anions and cations in the body compartments. Concentrations in mmol/L. Proteins and organic anions have not been taken into account.

most important intracellular cation, potassium is mainly responsible for the osmotic pressure of the cell, the membrane potential and the conduction of impulses in nerve and muscle cells.

The extracellular fluid provides the connection between cells and organs. As the cells can maintain a constant internal milieu with very slight fluctuations in the composition of the ECS, the volume, osmolality, pH and ionic composition of the ECS must be precisely regulated by the body.

The electrolytes of the extracellular space, ECS, show a characteristic concentration pattern. Blood plasma contains a high concentration of sodium and chloride while, by contrast, the concentrations of potassium, calcium or magnesium are comparatively low. The concentrations of extracellular sodium and chloride are kept constant within narrow limits. They account for the largest part, by far, of the osmotic pressure of this compartment.

The volumes and compositions of the ECS and ICS fluids are kept constant within narrow limits, a situation monitored by use of sensors located in the ECS. If they take part in the regulation of the osmotic pressure, we speak of osmoreceptors and in the event of blood volume or blood pressure we speak of volume and pressure receptors.

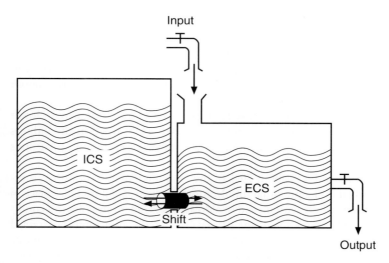

Fig. 3. Tank model of fluid spaces. ECS: Extracellular space. ICS: Intracellular space. Shift: Shift between ICS and ECS.

A steady state is reached when input and output (external balance) and the exchange between ICS and ECS (inner balance) are in equilibrium.

A decreased concentration of an electrolyte in the extracellular space can be caused by disturbance of the external and/or internal balance. A reduced input and/or increased output can cause a lowered concentration as well as a shift from extracellular to intracellular space (Fig. 3). An external imbalance leads to deficiency in the body with respect to the electrolyte, but an isolated internal imbalance does not. Similar considerations apply to increased electrolyte concentrations.

The electrolytes sodium and chloride possess relatively narrow serum reference intervals of 135–145 mmol/L for Na^+ and 98–106 mmol/L for Cl^-, compared to the relatively wide reference interval of 3.5–5.1 mmol/L for K^+ ions. Clinically relevant concentration changes of Na^+ and Cl^- therefore occur even with relatively small changes in body water.

1.2 Sodium

Physiology

The human body contains about 60 mmol sodium ions per kg body weight, i.e. of 70 kg body weight about 97 g (= 4200 mmol) are Na^+ ions. Approximately 40 mmol/kg (= 70%) are exchangeable. The rest, about 30%, is bound in skeleton.

The daily sodium intake is of the order of 150 mmol. With lower intake, the Na^+ ions are almost completely reabsorbed in the kidneys and, in the converse situation, there is increased excretion of Na^+ ions (Fig. 4).

In the kidneys about 70% of the total filtered amount of Na^+ is reabsorbed in the proximal tubule. More precise regulation takes place in the distal tubule with the aid of aldosterone. Under physiological conditions, only about 1% of the primarily filtered sodium ions is excreted in the final urine. In addition to glomerular filtration and aldosterone action, other physical phenomena and hormonal actions are discussed in connection with renal Na^+ excretion (e.g. atrial natriuretic peptide (ANP), brain natriuretic peptide (BNP)).

Electrolytes in the Serum

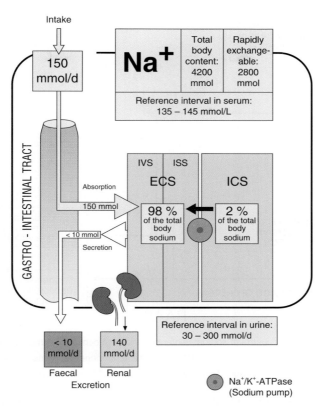

Fig. 4. Sodium balance. ECS: Extracellular space. ICS: Intracellular space. ISS: Interstitial space. IVS: Intravascular space.

The activity of the Na$^+$/K$^+$-ATPase in the cell membrane causes Na$^+$ to be removed from the cells and K$^+$ to be pumped into the cells. Cell membranes are not freely permeable to most substances. By contrast, water is almost freely permeable. This explains why the osmolalities of the intra- and extracellular spaces are identical in spite of their differing electrolyte composition and electrolyte concentrations.

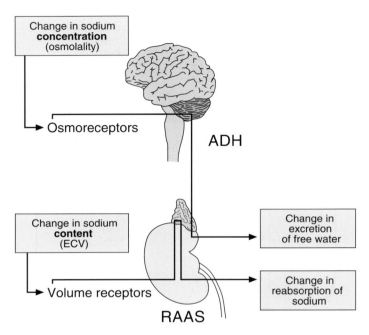

Fig. 5. Feedback control of sodium content and sodium concentration (osmolality) in the extracellular space. ADH: Antidiuretic hormone (vasopressin). ECV: Extracellular volume. RAAS: Renin-angiotensin aldosterone system.

- Deviations in the sodium content with intact osmoregulation lead to disorders of extracellular volume (ECV).
- Deviations in the water content lead to disorders of ECV and intracellular volume (ICV).

Even though not all of the details of extracellular volume regulation have been elucidated, the sodium and water balances of the body are essentially controlled by the following hormonal feedback mechanisms (Fig. 5).

1. The content of sodium ions in the ECV is mainly regulated via the renal excretion and reabsorption of Na^+ by means of the renin-angiotensin-aldosterone system. The regulatory parameter is the "effective" ECV. Volume and baroreceptors in the juxta-glomerular apparatus and baroreceptors in the carotid sinus and aortic arch are involved

as well as, for lower pressure changes, baroreceptors in the atria and major thoracic veins.
2. The sodium concentration is regulated via the renal excretion and reabsorption of free water with the aid of ADH and via the thirst mechanism. The regulatory parameter is the "effective" osmolality in the ECS. Osmoreceptors located in the hypothalamus are predominantly involved.

For more details on physiology see [111] and Seldin and Giebisch (1992) (rev. lit., p. 160).

In the serum, about 98.5% of the sodium ions are free, 1% is bound to proteins and 0.5% is bound e.g. to hydrogen carbonate (Fig. 6). The "active" fraction, however, is only about 75% of the total concentration as, in non-infinitely diluted solutions, electrostatic interactions increase in line with the quantity and charge of all the particles present in the

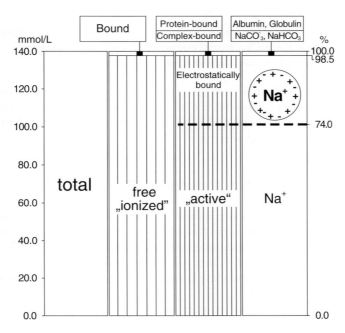

Fig. 6. Sodium fractions in serum (modified from [76]).

Na⁺ Concentration

		HYPERNATREMIA	NORMONATREMIA	HYPERNATREMIA
		Na⁺ concentration ↓ Free water Rel.* increased	Na⁺ concentration n Free water n	Na⁺ concentration ↑ Free water Rel.* reduced
Na⁺ content	HYPOVOLEMIA Na⁺ content ↓ ECV ↓	Severe Na⁺ deficiency Rel.* water excess **I** Na⁺ loss Renal, gastrointestinal	Na⁺ deficiency Water deficiency **II** Na⁺ loss Renal, gastrointestinal	Na⁺ deficiency Rel.* Water deficiency **III** Coma Diabeticum
	NORMOVOLEMIA Na⁺ content n ECV n	Normal Na⁺ content Rel.* water excess **IV** SIADH	Normal Na⁺ content Normal water content **V** Normal	Normal Na⁺ content Rel.* water excess **VI** Diabetes insipidus
	HYPERVOLEMIA Na⁺ content ↑ ECV ↑	Na⁺ excess Rel.* water excess **VII** Heart failure Liver failure Renal failure	Na⁺ excess Water excess **VIII** Edema	Na⁺ excess Rel.* water deficiency **IX** "Shipwreck"

Fig. 7. Disorders of sodium content and sodium concentration. ECV: Extracellular volume. SIADH: Syndrome of inadequate ADH secretion. ↑ increased, n normal, ↓ reduced, * relative in relation to sodium content.

solution, i.e. with ionic strength. The activity coefficient in such solutions is less than 1. The activity coefficient of sodium varies in the serum within narrow limits around 0.747, primarily because of the relatively constant composition of serum.

Pathophysiology and Therapy

The amount of sodium ions in the ECV and the concentration of sodium ions are regulated separately. Disorders in both systems can occur independently [108]. All combinations of normal or altered sodium content and normal or changed sodium concentration are possible (Fig. 7).

Disorders should therefore be considered separately for purposes of differential diagnosis.

Changes in sodium content are primarily diagnosed clinically (Fig. 8):

Electrolytes in the Serum 13

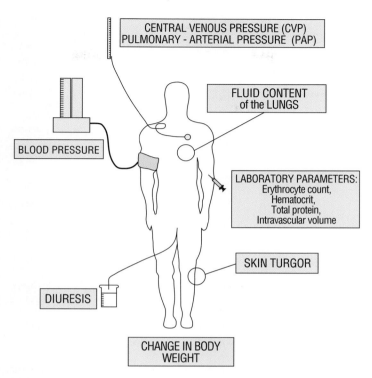

Fig. 8. Parameters for the assessment of sodium content in the extracellular space.

- Reduced sodium content: dehydration, reduced skin turgor, decreased pressure in the intravascular space (arterial pressure, venous pressure, central venous pressure, pulmonary arterial pressure, pulmonary capillary wedge pressure), reduced diuresis, reduction in body weight. Typically intravascular volume is reduced, whereas red blood cells, hematocrit and serum protein concentration are elevated.

- Increased sodium content: peripheral edema, pulmonary edema, increased pressure in intravascular space, increased weight. Intravascular volume is increased, whereas red blood cells, hematocrit and serum protein concentration are decreased.

Electrolytes in the Serum

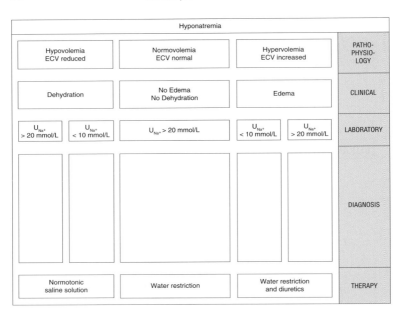

Fig. 9. Differential diagnosis and treatment of hyponatremia. ECV: Extracellular volume. U_{Na^+}: Sodium concentration in the urine.

Clinical chemistry parameters are of less informative value for the diagnosis of disorders of Na^+ content. Rough indications may be provided by the erythrocyte count, hematocrit, colloid osmotic pressure and protein concentration in the serum. The direct determination of intravascular volume (e.g. by means of radio-labeled albumin) is reserved for specific situations.

A decreased or increased sodium concentration in the serum is identified by the corresponding laboratory analysis.

Disorders of sodium content are treated by administration of physiological saline solution (Na^+ deficiency) or by withdrawal of sodium ions by diuresis (Na^+ excess).

Disorders of sodium concentration are treated by administration of Na^+-free water (increased Na^+ concentration) or by restriction of water (lowered Na^+ concentration) (Fig. 9).

Hyponatremia

Hyponatremia is said to be present if the sodium-ion concentration in the serum is less than 135 mmol/L [31]. Serum concentrations of less than 120 mmol/L are potentially life-threatening. Hyponatremia is one of the commonest electrolyte disorders.

Hyponatremia must be distinguished from pseudohyponatremia. The latter occurs when the water content of the serum is reduced by hyperlipoproteinemia or massive paraproteinemia (see Appendix, Tab. 49). Pseudohyponatremia does not occur if sodium determinations are performed by ion-selective electrodes without dilution of the sample.

Hyponatremia is not always equivalent to reduced osmolality. Glucose and other sugars (e.g. mannitol and sorbitol) can, if present in high concentrations in the serum, increase the osmolality of the serum and shift water out of the cells into the extracellular space: "water-shift" hyponatremia. The subsequent osmotic diuresis quickly balances this hyponatremia, allowing hypernatremia to develop.

The symptoms of hyponatremia are those of hypo-osmolality. They are based on the development of brain edema (cellular hyperhydration) and are expressed in cerebral symptoms: headaches, anorexia and nausea, confusional states, cerebral seizures, somnolence and even coma.

Hyponatremia with Normovolemia

Despite an increased water content there is no increase in renal water excretion in this disorder. In spite of hyponatremia and hypo-osmolality, unrestrained ADH secretion occurs. With intact volume regulation (recognizable by normovolemia and high sodium excretion in the urine) the ADH concentration remains inadequate (SIADH: syndrome of inadequate ADH secretion or Schwartz-Bartter syndrome; Tab. 1).

The differential diagnosis of hyponatremia is explained in Fig. 9.

The therapy of hyponatremia with normovolemia consists of the removal of excess water by restricting the intake of free water: thirsting, no water p.o. or i.v., possibly diuretics with consecutive equilibration of the diuresis with isotonic sodium chloride solution. Too rapid clinical correction involves the danger of central pontine myelinolysis. Only in an emergency with serum sodium concentrations of less than 120 mmol/L,

Table 1. Causes of hyponatremia with normovolemia

Syndrome of inadequate ADH secretion
(SIADH, Schwartz-Bartter syndrome)

• Tumors	Bronchial, pancreatic or prostatic carcinoma, lymphoma, thymoma, etc.
• Pulmonary diseases	Tuberculosis, aspergillosis, abscesses, pneumonia of other origin
• Cerebral diseases	Post-traumatic, hemorrhages, meningitis, encephalitis, etc.
• Stress	Emotional, postoperative, pains
• Endocrine causes	Hypothyroidism, panhypopituitarism, adrenal cortical insufficiency
• Drugs	Oxytocin, vincristine, cyclophosphamide, non-steroidal antiinflammatory drugs

might a more rapid correction of the hyponatremia with hypertonic sodium chloride solution be indicated:

5% sodium chloride solution i.v., infusion rate < 100 mL/h and < 400 mL/day.

Continuous monitoring of the serum sodium and serum osmolality is mandatory. The rise in serum sodium should not exceed 1 mmol/L per hour. Treatment of the primary disease (inflammation or tumor) is obvious in chronic primary hyponatremia. If this is not possible, then the best therapy at present is inhibition of the ADH secretion with demethylchlortetracycline at a dose of 600–1200 mg/day.

Hyponatremia with Hypovolemia

An extreme renal or extrarenal (e.g. gastrointestinal) loss of sodium leads to a reduction in the effective circulating volume with consecutive nonosmotic stimulation of ADH secretion. This is the only syndrome that shows hyponatremia with sodium deficiency, leading to considerable confusion between sodium concentration and sodium content (Tab. 2).

Therapy therefore involves physiological saline solution or Ringer lactate solution. In the presence of simultaneous acidosis, sodium bicarbonate may also be given.

Table 2. Causes of hyponatremia with hypovolemia

Renal sodium loss
- Salt-losing nephropathies
- Mineralcorticoid deficiency
- Diuretics

Extrarenal sodium loss
- Gastrointestinal loss (vomiting, diarrhea, drainage, fistula)
- Surface loss (burns, severe sweating)
- "Third-space" problems (fluid loss into third space)

As a rule volume deficiency with hyponatremia in adults generally occurs only above an ECV deficit of 3–6 L. With manifest clinical symptoms, therefore, at least this quantity of fluid must be infused.

Hyponatremia with Hypervolemia

In severe edematous diseases of cardiac, renal or hepatic origin, there may be, despite a massive absolute increase in ECV, a reduction in the effectively circulating blood volume. The deficiency in effective blood volume triggers ADH secretion via a non-osmotic stimulus. At the expense of osmoregulation an attempt is made to maintain effective volume. Despite the hyponatremia, therefore, free water continues to be re-absorbed renally and the extracellular fluid diluted (dilution hyponatremia). Despite a massive increase in total body sodium content, hyponatremia occurs. The typical example is severe congestive heart failure.

The treatment of dilution hyponatremia consists of the removal of excess Na^+ ions and free water. Administration of diuretics (in extreme cases also extracorporeal fluid withdrawal) as well as an improvement in the hemodynamic situation and renal perfusion (reduction in afterload, positive inotropic substances, dopamine) are needed to increase the excretion of sodium chloride. It is imperative that this treatment is combined with a reduction in excess free water by limiting intake. Strict stoppage of water intake p.o. or i.v.

As hyponatremia in heart failure develops slowly, clinical symptoms are very seldom attributable to hypo-osmolality. Sodium administration is therefore not only unnecessary, but even dangerous.

Hypernatremia

Hypernatremia is said to be present if the sodium ion concentration in the serum exceeds 145 mmol/L. Hypernatremia always means an increase in osmolality and a relative deficiency of free water. The body is protected from a deficiency of free water by two mechanisms. On the one hand, by the thirst mechanism and, on the other, by ADH-stimulated renal regulation of water excretion. For this reason, extreme hypernatremias are much rarer than hyponatremias. Severe hypernatremia arises only if normal or abnormal water losses are not adequately compensated via the thirst mechanism.

Differential diagnosis and treatment of hypernatremia are outlined in Fig. 10.

Hypernatremia leads, via increased osmolality in the ECS, to intracellular dehydration. The symptoms are restlessness and confusional states (delirium and hallucinations), cerebral seizures, lethargy, possibly coma. Neurological symptoms may also be caused by small intracerebral hemorrhages.

Hypernatremia				
Hypovolemia ECV reduced		Normovolemia ECV normal	Hypervolemia ECV increased	Pathophysiology
Dehydration		No edema	Edema	Clinical
U_{Na}^+ >20 mmol/L	U_{Na}^+ <20 mmol/L	U_{Na}^+ non-characteristic	U_{Na}^+ >20 mmol/L	Laboratory
Plasma osmolality >300 mosmol/kg				
Renal waterloss (osmotic diuresis)	Extrarenal water-loss (diarrhea, sweating)	Diabetes insipidus	Excessive Na^+ content	Diagnosis
Normotonic saline solution		Water administration	Water restriction and diuretics	Therapy

Fig. 10. Differential diagnosis and treatment of hypernatremia.
ECV: Extracellular volume. U_{Na^+}: Sodium concentration in the urine.

Table 3. Causes of hypernatremia with normovolemia

Loss of water
- Renal
  Diabetes insipidus (central, nephrogenic)
- Extrarenal
  Hyperventilation, tracheostomy, increased sweating (fever, increased temperature)

Inadequate intake of water
- Patient will not drink
  Hypodipsia (essential hypernatremia)
- Patient cannot drink
  Patient confined to bed, coma
- Patient may not drink
  Medically prescribed fluid withdrawal (e.g. pre-operative)

Hypernatremia with Normovolemia

This is caused by the loss or inadequate intake of water without simultaneous loss of sodium ions and other electrolytes (Tab. 3). The patients are considered to be clinically normovolemic if there are no massive water deficits [48].

Therapy

- Removing cause (e.g. ADH deficiency).
- Supply of (sodium-free) water with monitoring of serum osmolality and serum sodium.

For prophylaxis in severely ill patients, those confined to bed as well as pre- and postoperative patients, adequate fluid intake must be ensured. The disorder is corrected by administration of oral fluids (mineral water, tea). Only if this is not possible should an isotonic 5% glucose solution be infused (distilled water should not be infused as it causes hemolysis!) The first liter of 5% glucose solution is infused in the first 4 hours and the first half of the calculated water deficit given in the first 24 hours. Further correction is based on the clinical findings. Serum sodium and osmolality values are used for monitoring.

Too rapid correction involves the danger of cerebral edema; in extreme cases central pontine myelinolysis may also develop.

Hypernatremia with Hypovolemia

This situation occurs when water loss is accompanied by a relatively lower loss of sodium ions (Tab. 4). If the water loss is greater than 2% of the body weight, the thirst mechanism is triggered. If it is greater than 6%, clinical symptoms appear. An extreme water loss of up to 10% can arise in diabetic metabolic derangement. In this case other electrolytes in addition to sodium ions are lost through diuresis. In extreme cases, the loss of sodium ions can be as much as 15% of the total body sodium.

Table 4. Causes of hypernatremia with hypovolemia

Renal loss of water with low content of electrolytes
• Osmotic diuresis in diabetic metabolic derangement

Extrarenal loss with low content of electrolytes
• Hypotonic loss of fluid in diarrhea, burns, heat stroke

The hyperosmolality caused by hyperglycemia withdraws water from the ICS, which results in hyponatremia. For every 100 mg/dL (5.6 mmol/L) increase in glucose, there is a fall in serum sodium of 1.6 mmol/L. As hyperglycemia progresses, osmotic diuresis develops and the water loss thus induced generally causes hypernatremia.

To compensate for the existing deficits of water and sodium, therapy starts with administration of about 1 L physiological saline solution to adjust volume, followed by hypotonic (e.g. 0.45%) NaCl solution to reduce osmolality. Simultaneously, insulin deficit is treated (and, where appropriate, potassium and magnesium replaced).

Hypernatremia with Hypervolemia

The increase in sodium content with a relative water deficit is almost always attributable to an "accident". If shipwrecked persons drink large quantities of seawater (sodium concentration about 450 mmol/L), hyperosmolality develops. If the feeling of thirst induces more drinking, then this leads to hypervolemia since the maximum excretion capacity for sodium is about 1500 mmol/day.

On administration of hypertonic salt solution to children or to patients with renal failure a much feared therapeutic "shipwreck" may develop, as can also occur with sodium bicarbonate therapy during resuscitation (Tab. 5).

Table 5. Causes of hypernatremia with hypervolemia

Excessive sodium ion intake

• Sodium chloride or sodium bicarbonate infusion ("therapeutic shipwreck")
• Drinking of seawater (actual shipwreck)

Therapeutically, reduction of osmolality is achieved by administration of low sodium solutions (5% glucose solution) and simultaneous removal of excess sodium chloride with diuretics. As patients with renal failure are often involved, the excess water sometimes has to be removed by extracorporeal means.

Normonatremia with Hypovolemia

The equally large loss of sodium ions and water – with reference to extracellular concentration – leads to hypovolemia with normonatremia. It occurs in gastrointestinal disorders (e.g. vomiting, diarrhea and fistulae) and may occasionally be caused renally (polyuric nephropathy) or iatrogenically (e.g. ascites puncture). It can lead to renal failure and, if onset is rapid, to hypovolemic shock. By definition the sodium concentration in the plasma in this situation is normal, so that the diagnosis can be made solely on the basis of clinical findings. For therapy administer isotonic Ringer lactate solution or plasma expander in isotonic saline.

Normonatremia with Hypervolemia

Equal increases in water and sodium ions – relative to extracellular concentration – lead to hypovolemia with normonatremia and to increased weight and edema. Primary edema occurs in renal insufficiency if the intake exceeds the renal excretion capacity. Secondary edema develops in heart failure and hypoproteinemia (e.g. nephrotic syndrome, liver cirrhosis) as a result of the reduction in effective arterial blood volume. The diagnosis is essentially based on clinical criteria. Therapy consists of restriction of water and sodium chloride intake as well as acceleration of renal excretion by administration of diuretics, taking into account the underlying disease.

1.3 Chloride

Physiology

A body weighing 70 kg contains approximately 80 g (= 2260 mmol) of chloride ions. About 30% are fixed, primarily in the erythrocytes and muscle cells, and only about 60 g (1600 mmol) of chloride ions are rapidly exchangeable.

Of the approximately 80 g of chloride ions in the human body, about 88% are located in the ECS (32% in the skeleton) and about 12% in the ICS (Fig. 11). The daily chloride requirement is about 5 g (150 mmol). An almost complete absorption takes place in the small intestine. Up to 99% of chloride ions are reabsorbed in the kidney from the primary urine.

Chloride is the most important concomitant ion to sodium ions in the extracellular space. It is the commonest routinely determined anion. Other important anions such as bicarbonate and the anions of organic acids are generally less frequently determined. Their total concentration is reflected in the "anion gap" (Fig. 12).

In the majority of cases, chloride concentration and sodium concentration behave coordinately, e.g. a 10% reduction in sodium concentration should be accompanied by a 10% reduction in chloride concentration. If the sodium and chloride concentrations do not change proportionally, an acid-base balance disorder may be suspected.

Chloride and sodium concentrations in sweat are elevated in case of mucoviscidosis.

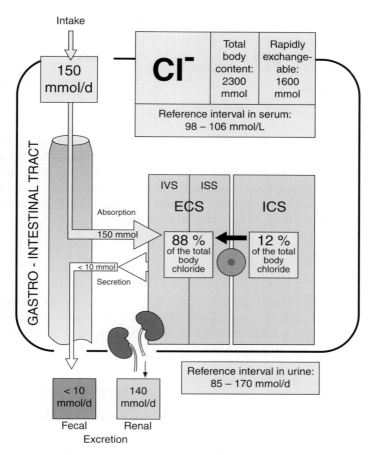

Fig. 11. Chloride balance. ECS: Extracellular space. ICS: Intracellular space. ISS: Interstitial space. IVS: Intravascular space.

Pathophysiology and Therapy

Hypochloremia

Hypochloremia is said to be present if the serum chloride concentration is less than 98 mmol/L. Relative hypochloremia is characterized by a simultaneously raised bicarbonate concentration, which is observed in metabolic alkalosis and chronic respiratory acidosis (Tab. 6).

Table 6. Causes of relative hypochloremia

Metabolic alkalosis
Chronic compensated respiratory acidosis

Serum chloride determination plays a specific role in the differential diagnosis of hypercalcemia (Tab. 19). Hyperparathyroidism has a tendency to produce hyperchloremic metabolic acidosis as the result of renal bicarbonate loss. In tumor-induced hypercalcemia there is, by contrast, a tendency to metabolic alkalosis and therefore to hypochloremia (treatment of hypochloremia: see hyponatremia).

Hyperchloremia

Hyperchloremia exists when the serum chloride concentration exceeds 106 mmol/L. Relative hyperchloremia is said to be present if there is simultaneously reduced bicarbonate concentration. This is observed mainly in metabolic acidosis (Tab. 7, Fig. 12).

Table 7. Causes of relative hyperchloremia

Hyperchloremic metabolic acidosis
• Administration of hydrochloric acid or equivalents (hydrochloric acid or ammonium chloride, arginine hydrochloride, lysine hydrochloride)
• Bicarbonate losses
  Renal bicarbonate losses (proximal tubular acidosis, distal tubular acidosis)
  Gastrointestinal bicarbonate losses (diarrhea, etc., uretero-sigmoidostomy)
  Hyperparathyroidism
Chronic respiratory alkalosis
• Hyperventilation

When replacement is carried out with hydrochloric acid or equivalents, e.g. the hydrochlorides of the amino acids arginine or lysine, or in renal or extrarenal bicarbonate loss, one sees hyperchloremic acidosis with relative hyperchloremia.

Relative hyperchloremia with lowered bicarbonate concentration is also observed in chronic respiratory alkalosis.

1.4 Anion Gap in the Serum

Physiology

For reasons of electroneutrality, a given number of Na^+ cations in the ECS must be matched by a corresponding number of anions. The anions of the ECS are mainly chlorides and bicarbonates. The sum of both anion concentrations is, however, less than that of the sodium ion concentration. The difference between the serum concentrations of Na^+ on the one hand and Cl^- and HCO_3^- on the other is known as the anion gap (AG).

$$\text{Anion gap in the serum} = Na^+ - [Cl^- + HCO_3^-]$$

Concentrations in mmol/L.

The anion gap arises from the fact that the sum of the concentrations of the anions not taken into account in the anion gap (e.g. proteinates, phosphates, lactate, pyruvate, etc.) is greater than the sum of ignored cations (K^+, Ca^{2+}, Mg^{2+}, see Fig. 12).

The anion gap is normally 8–16 mmol/L. It is widened, for example, in metabolic acidosis, in which the bicarbonate concentration in the serum is lowered with a concomitant increase in the concentration of anionic organic acids, e.g. lactate, acetoacetate, β-hydroxybutyrate, salicylate, formate (in methanol poisoning) and oxalate and glycolate (in ethylene glycol poisoning). It is reduced in extreme hypercalcemia, hypermagnesemia and lithium intoxication.

If the sodium concentration is determined in serum water (by ISE) and the total chloride and bicarbonate concentrations are determined in serum the following must be borne in mind:

1. Normal water concentration of the serum:
   The anion gap is approx. 12–20 mmol/L since the sodium concentration in serum water is about 3% higher than in serum. On the other hand, if the concentration is given as "ionized sodium" the reference interval for the anion gap remains 8–16 mmol/L.

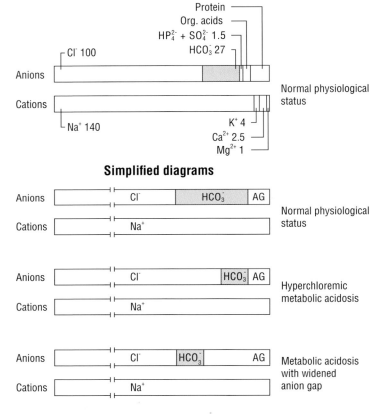

Fig. 12. Anion gap (AG) in the serum. Data in mmol/L.

2. Water concentration of the serum reduced (e.g. hyperlipemia):
   The anion gap increases to the extent that the water concentration decreases. It increases to above 12–20 mmol/L if the sodium concentration in serum water is given and to above 8–16 mmol/L if the concentration of "ionized sodium" is given. The reason for the increase in the anion gap is the decrease in the chloride and bicarbonate concentrations.
3. Water concentration of the serum increased (e.g. hypoproteinemia):
   The anion gap decreases to the extent that the water concentration increases. It decreases to below 12–20 mmol/L if the sodium con-

centration in serum water is given and to below 8–16 mmol/L if the concentration of "ionized sodium" is given. The reason for the decrease in the anion gap is the increase in the chloride and bicarbonate concentrations.

For calculation of the anion gap it is therefore advisable to use anion concentrations from one system only, e.g. only serum or only serum water (e.g. ionized sodium, ionized chloride). In this case the anion gap is independent of the water content of the serum.

Pathophysiology

The cardinal finding in metabolic acidosis is, in addition to pH lowering, reduced bicarbonate concentration. The reduced serum bicarbonate is replaced by the anions that are not normally measured in routine examinations (high anion gap) or by chloride (normal anion gap). Metabolic acidoses can thus be subdivided into:

- Metabolic acidosis with widened anion gap.
- Hyperchloremic metabolic acidosis (with normal anion gap, see page 56).

Causes of metabolic acidosis with widened anion gap (Tab. 8) are always the result of endogenous or exogenous input of acid ions. The seven most important causes can be summarized under the acronym "KUSMALE":

K: Ketoacidosis (ketone bodies).
U: Uremia (retention of anionic non-volatile acids).
S: Salicylate poisoning.
M: Methanol poisoning (formate).
A: Aethanol poisoning (lactate, keto acids).
L: Lactic acidosis (lactate).
E: Ethylene glycol poisoning (oxalate, glycolate).

Table 8. Metabolic acidosis with widened anion gap

- Ketoacidosis: diabetic, alcoholic, starving
- Renal failure
- Lactic acidosis
- Intoxication with salicylates, methanol, alcohol, ethylene glycol and paraldehyde

1.5 Osmolality

Physiology

Osmolality represents the total number of all particles dissolved in 1 kg of solvent (e.g. water). The measurements are carried out mostly in serum/plasma or urine.

The plasma osmolality is the most important factor for assessing internal water balance. The diffusion of water from compartments of lower osmolality to compartments of higher osmolality continues until equilibrium is achieved. Regulation takes place via the thirst center and via secretion of antidiuretic hormone.

Pathophysiology

Plasma osmolality depends primarily on the concentration of sodium, chloride and bicarbonate ions plus glucose and urea which, in diabetes mellitus or renal failure, can achieve osmotically significant concentrations. The most important exogenous compound is ethanol. Even a moderately increased blood alcohol concentration leads to a marked rise in osmolality in the serum.

The osmolality of the serum can be estimated using the following empirical equation:

$$\text{Serum osmolality [mosmol/kg]} = 1.86 \times [Na^+] + [\text{urea}] + [\text{glucose}] + 9$$

Concentration data in mmol/L.
The factor 1.86 takes into account the concentrations of Cl^- and HCO_3^-.

According to the Système International d'Unités (SI) mosmol/kg should be replaced by mmol/kg (without change of numerical value).

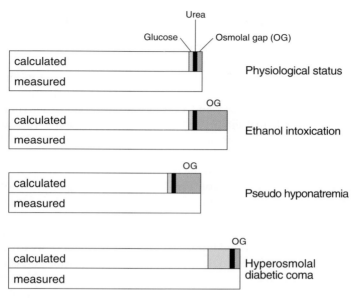

Fig. 13. Osmolal gap (OG).

The difference between the measured and the calculated osmolality is the "osmolal gap" (Fig. 13). The osmolal gap is always positive and may be as high as +10 mosmol/kg in healthy subjects (when using this equation). It is caused by osmotically active particles that are not taken into account in the calculation. The osmolal gap is widened when the particles that are not taken into account in the calculation of osmolality are present in higher concentrations, e.g. lactate. With a blood ethanol concentration of 1 g/kg the measured serum osmolality increases, for example, by 29 mosmol/kg. Similarly, the omolal gap increase with, for example, infusions of mannitol. In pseudohyponatremia the osmolal gap is widened as, for example, hyperlipidemia decreases the concentration of analytes in serum, but not the osmolality which is related to serum water.

1.6 Potassium

Physiology

Clinically, the importance of potassium is based primarily on its effect on neuromuscular excitability. The potassium concentration gradient between ICS (160 mmol/L) and ECS (4 mmol/L) mainly determines the electrical potential difference across the cell membrane. Increases or reductions in total potassium content initiate neuromuscular effects which in extreme cases – mainly when the cardiac muscle is involved as the target organ – can threaten the patient's life.

A body weighing 70 kg contains 140 g (= 3600 mmol) of K^+ ions. About 20% are fixed in the muscles and about 2800 mol K^+ ions are rapidly exchangeable.

Potassium is the main cation in the ICS (Fig. 14). About 98% of total body potassium is found intracellularly and only about 2% extracellularly. In contrast, for example, with plasma sodium, plasma potassium concentration in the extracellular space correlates at normal "internal balance", i.e. especially with normal pH, sufficiently well in clinical respects with total body potassium [115].

As a rule of thumb it may be stated that, in hypokalemia, for each 1 mmol/L reduction in plasma concentration, there is a total potassium deficit of about 200–400 mmol. In hyperkalemia, for each 1 mmol/L plasma potassium increase there is an excess of about 100–200 mmol in total potassium content.

Differences between potassium concentration and potassium content arise primarily in disorders of the acid-base balance. With normal potassium content the plasma potassium concentration can fluctuate between 6.5 mmol/L at a pH of 7.0 and 2.5 mmol/L at a pH of 7.7. As a rough estimate, a pH shift of 0.1 pH units produces a shift in plasma potassium concentration of 0.5 mmol/L (Fig. 15).

The human body takes in about 75 mmol of potassium ions daily with the food. This is normally absorbed 100% in the gastrointestinal tract. Only about 5 mmol are secreted into the intestinal tract and excreted fecally.

Potassium balance is regulated by the kidney. In the presence of an overload, the excess potassium is excreted via distal tubular secretion (up to 400 mmol K^+ ions per day). In cases of potassium deficiency, the

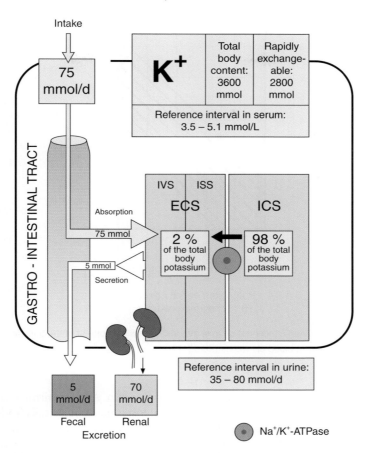

Fig. 14. Potassium balance. ECS: Extracellular space. ICS: Intracellular space. ISS: Interstitial space. IVS: Intravascular space.

excretion into the urine is reduced (< 10 mmol K$^+$ ions per day). Renal potassium regulation is effective down to a very low glomerular filtration rate.

To understand drug-induced potassium imbalance it is important to know that the distal tubular cells primarily secrete potassium, a process induced by aldosterone. Aldosterone secretion is stimulated pri-

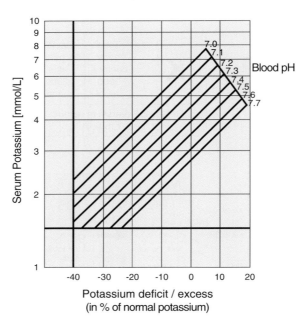

Fig. 15. Potassium concentration in serum and potassium content of the body at different blood pH values (from [111]).

marily by angiotensin II and this in turn by renin. Any impairment of the reaction chain renin-angiotensin II-aldosterone-tubular cells leads to disruption of the potassium balance [111]. Catecholamines stimulate renin secretion; they promote renal potassium excretion via increased aldosterone secretion. In addition, potassium uptake into the cells is promoted by catecholamines.

Pathophysiology and Therapy

Hypokalemia

Hypokalemia is said to be present if the plasma potassium value falls below 3.5 mmol/L.

Table 9. Causes of hypokalemia

Negative external balance
Reduced intake
• Low potassium diet • Diarrhea, other gastrointestinal malabsorption disorders, laxative abuse
Increased potassium loss

• Renal loss
  Renal tubular disorders
  Polyuria
  Diuretics

• Hypermineralocorticoid states
  With raised renin level
    Secondary hyperaldosteronism
    Bartter's syndrome
    Pseudo-Bartter's syndrome (abuse of diuretics and laxatives, chronic vomiting)
    Renal artery stenosis
  With decreased renin level
    Primary hyperaldosteronism
    Enzyme defects of the adrenals (e.g. 17α-hydroxylase deficiency)
    Administration of substances with mineralocorticoid activity (liquorice, etc.)

• Extrarenal loss   Diarrhea and other gastrointestinal loss
Disorders of internal balance (potassium shift from extracellular to intracellular space)
• Metabolic alkalosis • Insulin • β-adrenergic action • Familial hypokalemic periodic paralysis

As described in the tank model (Fig. 3) hypokalemia can be caused by reduced input or raised output (negative external potassium balance). In addition it may be caused by a shift of extracellular potassium ions into the intracellular space (disorder of internal balance, see Fig. 16). If there is no metabolic alkalosis and no other cause for a potassium shift, then hypokalemia is attributable to potassium deficit (Tab. 9).

The main symptoms of potassium deficiency involve the skeletal muscles, with muscular weakness, loss of tone, extending to atonic paralysis, respiratory paralysis and rhabdomyolysis. Impairment of smooth muscle leads to intestinal atony and even paralytic ileus. The

Electrolytes in the Serum

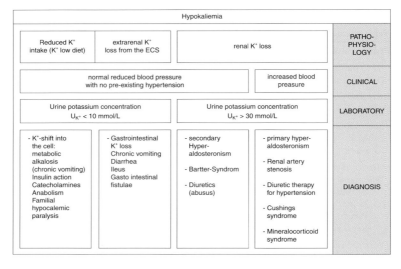

Fig. 16. Differential diagnosis of hypokalemia. ECS: Extracellular space. U_K^+: Potassium concentration in the urine.

most life-threatening effects are myocardial symptoms, with tachyarrhythmias and ECG changes (U-wave, T/U-combined wave, QT prolongation). The inhibiting action of cardiac glycosides on the Na^+/K^+ pump is intensified by hypokalemia.

In replacement therapy with potassium, the administration of large quantities of potassium should be avoided because of the potential fatal consequences for the heart. The oral route is the least dangerous. Intravenous therapy should be limited to acute symptomatic hypokalemia and to gastrointestinally induced hypokalemia.

There are various rules for correct procedure, one of which is known as the "rule of 40": The potassium concentration in the replacement solution should not exceed 40 mmol/L and the infusion rate should not exceed 40 mmol/h or 240 mmol/d. Naturally this rule should only be understood as a rough guide.

The following is recommended for the treatment of hypokalemia:

1. Increase potassium intake
   - In non-threatening hypokalemia ($K^+ > 3.0$ mmol/L, no cardiac symptoms, patient not taking digitalis) oral potassium replacement, e.g. potassium-rich food, KCl p.o.
   - In life-threatening hypokalemia and/or inadequate gastrointestinal absorption intravenous potassium administration ("40 rule", see above).
2. Potassium-sparing diuretics
   - Spironolactone
   - Triamterene
   - Amiloride

Hyperkalemia

Hyperkalemia is said to be present if the serum potassium ion concentration is increased to > 5.1 mmol/L (Fig. 17). Life-threatening symp-

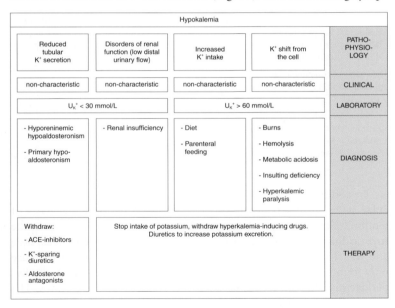

Fig. 17. Differential diagnosis and therapy of hyperkalemia. U_{K^+}: Potassium concentration in the urine. ACE: Angiotensin converting enzyme.

toms occur mostly at serum concentrations > 7 mmol/L. Extreme K⁺ values (> 10 mmol/L) are not compatible with life. In such cases a sampling error (e.g. collection during running K⁺ solution infusion or hemolysis) or incorrect storage (e.g. storage of whole blood in a refrigerator) should be suspected (Tab. 10).

Table 10. Causes of hyperkalemia

Positive external balance
Increased intake:
• Abnormally high input (parenteral)
Reduced excretion:
• Reduced renal output Reduced urinary flow (oliguria, severe renal insufficiency) Reduced potassium secretion in distal tubule
Reduced response to aldosterone (pseudohypoaldosteronism) Reduced aldosterone action, adrenal insufficiency Hypoaldosteronism Reduced angiotensin II production, angiotensin-converting enzyme (ACE) inhibitors Reduced renin production: diabetic nephropathy, hyporeninemic hypoaldosteronism, beta-receptor blockers, non-steroidal anti-inflammatory drugs Potassium-sparing diuretics (amiloride, triamterene, spironolactone)
Disorders of internal balance (potassium shift from intracellular to extracellular space)
• Acidosis • Cell lysis, rhabdomyolysis, hemolysis • Familial hyperkalemic paralysis • Malignant hyperthermia • Succinylcholine therapy

In addition to non-characteristic general symptoms, certain muscular states in hyperkalemia resemble those in hypokalemia: Loss of tendon reflex, atonic paralysis extending to respiratory paralysis. In the myocardium hyperkalemia leads to bradyarrhythmias and even asystole and ventricular fibrillation. The characteristic high-peaked T-waves are only found with previously normal ECGs. With prior T-negativity hyperkalemia can lead to a "pseudo"-normalization of the T-waves. The action of cardiac glycosides is attenuated in hyperkalemia.

The first step in the treatment of hyperkalemia consists of stopping potassium intake (via fruits, e.g. bananas or K⁺-containing drugs, e.g.

potassium penicillin). Drugs (e.g. β-blockers, ACE inhibitors, aldosterone antagonists and K^+-sparing diuretics) which may cause hyperkalemia should be withdrawn. If necessary, the next step is to bind the potassium ions in the gastrointestinal tract (administration of ion-exchange resins). In acute emergencies one should attempt to shift the potassium ions into the intracellular space, either by bicarbonate therapy, insulin-glucose therapy or with β-mimetics. Calcium salts can protect the cell membrane against the actions of hyperkalemia.

Renal potassium excretion can be promoted by increasing diuresis (furosemide). If the renal elimination of potassium ions is not sufficient, then removal by hemodialysis should be considered.

Table 11. Therapy of life-threatening hyperkalemia

1st stage	Antagonism of hyperkalemic action on the membrane potential.
	• Calcium gluconate 10%, 2–3 ampoules of 10 mL i.v. Onset of action after 5 min, duration of action no more than 30 min.
2nd stage	Shift of potassium ions into the intracellular space.
	• Insulin/glucose infusion: 500 mL of 10% glucose solution with 10 U of regular insulin. Onset of action after 30 min, duration of action: several hours.
	• Sodium bicarbonate (especially in the presence of concomitant metabolic acidosis): 100 mmol of bicarbonate to insulin/glucose solution, subsequently 500 mL of 5% glucose solution.
	• β-Adrenergic agents
3rd stage	Extracorporeal removal of potassium ions.
	• Hemodialysis

With chronic hyperkalemia with values of < 7 mmol/L one option is the immediate suppression of all potassium intake and the administration of loop diuretics. Enterally administered exchange resins replace potassium ions either by sodium ions (Resonium) or by calcium ions (calcium polysterol) and are well-tried adjuvant drugs for hyperkalemia.

Life-threatening hyperkalemias greater than 7 mmol/L, especially in acute cases, require "3-stage emergency therapy" (Tab. 11).

1.7 Magnesium

Physiology

Next to potassium, magnesium is the most important intracellular cation. Many enzymes require magnesium ions for their catalytic action. Thus, for example, all ATP-dependent enzymatic reactions require Mg^{2+} as a cofactor in the ATP-magnesium complex. Mg^{2+} activates the Na^+/K^+-ATPase at the cell membrane. Adenylate cyclase, which catalyzes the formation of the second messenger cyclic AMP, is magnesium-dependent. A close interaction exists with the calcium balance. Magnesium is considered to be a physiological calcium antagonist.

Of a 70 kg total body weight, approximately 24 g, i.e. about 1000 mmol, are Mg^{2+} ions. About 69% of this is stored in bone and only 300 mmol Mg^{2+} are rapidly exchangeable. As a predominantly intracellular cation, magnesium has similarities with potassium, the majority is present in muscle cells (Fig. 18).

The magnesium requirement is approximately 15 mmol/d and this is provided by food. About 50% is absorbed in the small intestine.

Like calcium, about 70% of the magnesium ions are present in the serum in free form and 30% are bound, of which 25% are bound to proteins (especially albumin), the rest to citrate, phosphate and other complex formers (Fig. 19).

A range of 0.70–1.05 mmol/L is often given as the reference interval for magnesium. There is evidence that the lower limit of the reference range in fact reflects a widespread low dietary intake of magnesium. Experts believe that at a concentration of 0.70 mmol/L there is already a magnesium deficiency. On the basis of relevant studies they recommend that 0.75 mol/L should be taken as the lower limit [2, 124]. Moreover, studies show that even within this recommended reference interval there is a negative correlation between cardiac disturbances, e.g. arrhythmia, and the magnesium concentration.

Regulation of the magnesium concentration in the plasma takes place chiefly via the kidneys, here particularly via the ascending loop of Henle. Normally about 5% of the magnesium undergoing glomerular filtration is eliminated with the urine. With diminished magnesium intake the fractional excretion can fall to 0.5%, with Mg overload it can increase to >50%.

Electrolytes in the Serum

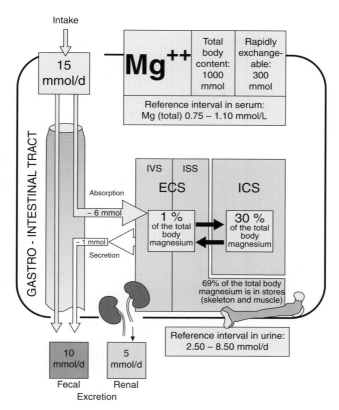

Fig. 18. Magnesium balance. ECS: Extracellular space.
ICS: Intracellular space. ISS: Interstitial space. IVS: Intravascular space.

The Mg^{2+} serum level is kept constant within very narrow limits between 0.65 and 1.05 mmol/L. Regulation takes place mainly via the kidneys, especially via the ascending loop of Henle. Normally about 5% of the glomerularly filtered magnesium is excreted in the final urine. Fractional excretion can fall during reduced magnesium intake to about 0.5% and with Mg overload increase to > 50%.

The precise regulatory mechanisms are not yet known, but there is apparently no one main factor responsible for Mg^{2+} excretion. Tubular reabsorption of magnesium is decreased in the event of hypermagnese-

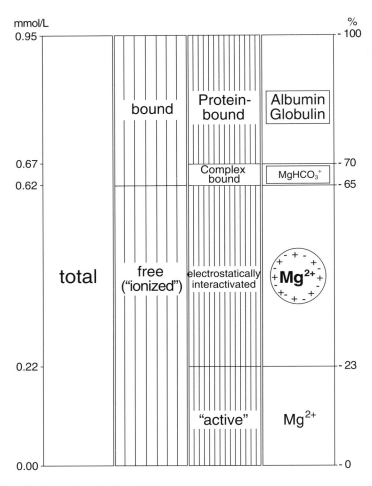

Fig. 19. Magnesium fractions in the plasma.

mia, hypercalcemia and increased natriuresis (due to e.g. hypervolemia, acetazolamide, thiazides, furosemide, osmotic diuresis). Tubular reabsorption of magnesium is increased in the event of hypomagnesemia, hypocalcemia, hypovolemia, therapy with K^+-sparing diuretics and increased secretion of several peptide hormones (growth hormone, calcitonin). PTH also leads to increased tubular reabsorption of magnesium.

Table 12. Symptoms of magnesium deficiency

Hypocalcemia-like symptoms
Spasmophilia, calf cramps, tetany, muscular weakness, dysphagia, esophageal spasm, intestinal cramps (refractory to calcium replacement).

Hypokalemia-like symptoms
Heart failure, angina pectoris (coronary disease, myocardial infarction), ECG changes (QRS prolongation, T-wave flattening, U-waves), arrhythmias (supraventricular tachycardia, ventricular extrasystoles, ventricular fibrillation).

Cerebral symptoms
Apathy, depression – but also increased excitability, confusion, delirium, epileptiform seizures, vertigo, nystagmus.

Due to concomitant hypercalcemia, however, hypomagnesemia is often observed in hyperparathyroidism.

Pathophysiology and Therapy

Hypomagnesemia

Because of the interactions between magnesium, potassium and calcium, hypomagnesemia exhibits symptoms which are similar to those of hypocalcemia or hypokalemia (Tab. 12). Because of the target organ resistance for parathyroid hormone in magnesium deficiency, hypocalcemia is, in fact, present in about 50% of cases. On the other hand, the hypomagnesemia symptoms resemble those of potassium deficiency (e.g. cardiac symptoms,
ECG changes, arrhythmias). As Mg^{2+} activates the Na^+/K^+-ATPase of the cell membrane, intracellular potassium deficiency occurs in the presence of magnesium deficit, and as a result of increased renal potassium loss there is also true hypokalemia in 50% of the cases. The action of cardiac glycosides is intensified by hypomagnesemia [29].

Magnesium deficiency states play an important role clinically. Specific causes of magnesium deficiency states can include lack of intake, but also gastrointestinal and renal losses or shifts of Mg^{2+} ions from the ECS into the ICS (Fig. 20, Tab. 13). Frequently, the diagnosis and differentiation of deficiency states are not conclusive from determination of the plasma magnesium alone. This differential diagnostic problem requires further clarification. The recent ability to determine free ("ionized") magnesium could be an important step in this direction. It per-

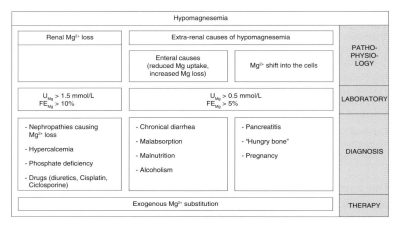

Fig. 20. Differential diagnosis and treatment of hypomagnesemia. U_{Mg}^{2+}: Magnesium concentration in the urine.

mits the recognition of pseudo-hypomagnesemia, which is caused by a reduction of the protein-bound fraction in hypoproteinemia. With citrate administration in the course of blood transfusion (e.g. liver transplantation) the free fraction is lowered with "normal" total concentration and increased complex bound fraction.

Further indications of a masked magnesium deficiency may be provided by the magnesium loading test (Fig. 21). The magnesium loading

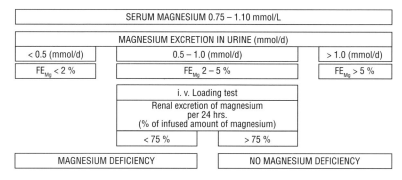

Fig. 21. i.v. Magnesium-loading test in suspected magnesium deficiency with normal serum magnesium concentration (according to [29]).

Table 13. Causes of hypomagnesemia

Negative external balance

Reduced intake
- Malnutrition
- Alcoholism
- Magnesium-free parenteral feeding

Increased loss
- Renal loss
  Diuretic phase after acute kidney failure, postobstructive diuresis, polyuria
  Congenital tubular defects
  Conn's syndrome (primary hyperaldosteronism), SIADH (syndrome of inadequate secretion of ADH)
  Hypercalcemia (including primary hyperparathyroidism), hyperthyroidism, thyroid hormone therapy
  Metabolic acidosis
  Diuretics
  Drugs (aminoglycosides, amphotericin B, cisplatin, ciclosporine and D-penicillamine)
- Gastrointestinal loss
  Chronic diarrhea (Crohn's disease, ulcerative colitis)
  Malabsorption (celiac disease and tropical sprue, pancreatogenic)
  Resection of the small intestine

Disorders of internal balance
(magnesium shift from the extracellular space)
- "Hungry bone" syndrome after total parathyroidectomy
- Acute pancreatitis
- Insulin administration
- Pregnancy

test is contraindicated in renal failure, hypermagnesemia and hypercalcemia. Magnesium-containing preparations as well as laxatives, antacids and diuretics must be withdrawn at least 48 h before the test.

Treatment of Hypomagnesemia

If there are clinical signs of magnesium deficiency (Mg^{2+} in the serum < 0.65 mmol/L), then the Mg^{2+} deficit is about 0.5 mmol/kg body weight. Dietetic correction can take place via the diet, e.g. with citrus fruits or magnesium salts. In acute situations or as an additive to parenteral nutrition, magnesium is infused. As, in most cases, simultaneous potassium deficiency is present, a potassium magnesium aspartate infusion solution is adequate.

- Oral therapy
  - Dietetic Mg^{2+} administration (citrus fruits)
  - Mg^{2+} salts in tablets or as soluble granules (Mg^{2+} content between 4 and 15 mmol per dose). Recommended intake for chronic magnesium deficiency: about 15 mmol/d.
- Intravenous therapy
  - Magnesium gluconate (1 ampoule of 2.4 mmol Mg^{2+}). Recommended intake: 0.25 mmol/kg body weight per day (up to 40 mmol/d) in 2000 mL 5% glucose solution over 8 h i.v.

Hypermagnesemia

Hypermagnesemia (> 1.2 mmol/L) and excess magnesium are generally life-threatening only in renal failure and thus have similar causes and sometimes similar symptoms as hyperkalemia (Tab. 14).

Clinical symptoms of hypermagnesemia are rarely marked or characteristic. They can be subdivided into neuromuscular symptoms (muscle weakness, paresis, central nervous sedation, respiratory paralysis, somnolence, coma), cardiovascular symptoms (hypotension, brady-

Table 14. Causes of hypermagnesemia

Positive external balance

Increased intake

- Parenteral: Magnesium therapy
- Enteral: Magnesium-containing antacids and laxatives
- Rectal: Magnesium-containing enemas

Reduced renal excretion

- Acute and chronic renal failure

Disorders of internal balance

Magnesium release from intracellular space

- Rhabdomyolysis
- Cell lysis after cytotoxic therapy
- Burns
- Trauma

cardia, ECG changes – similar to hyperkalemia) and general symptoms (nausea, vomiting).

Effective treatment of magnesium excess is stoppage of exogenous magnesium intake. In addition, calcium gluconate can be administered i.v. In chronic renal failure dialysis procedures must be used.

1.8 Calcium

Physiology

The calcium content of an adult is somewhat over 1 kg (25,000 mmol), i.e. about 2% of the body weight. Almost all the calcium (99%) is found in bones and only less than 1% is found in extraosseous ECS or in the ICS. The calcium content of the ECS is in dynamic equilibrium with the rapidly exchangeable fraction of the bone calcium, which is about 100 mmol (Fig. 22).

Calcium plays an essential role in many cell functions, e.g. in myocardial contractility, nerve conduction, hormone secretion and action and in various enzymatic reactions such as, for example, blood clotting. It is essential for bone mineralization [39].

The daily uptake of calcium is about 25 mmol (= 1 g), mainly from milk and milk products.

The intestinal intake normally amounts to 10 mmol/d and is dependent on vitamin D. There is also considerable intestinal secretion of calcium.

250 mmol calcium are glomerularly filtered per day (about 170 L primary urine with a calcium concentration of about 1.5 mmol/L). About 90% of the filtered calcium is reabsorbed, together with sodium, in the proximal tubule and the loop of Henle. The fine adjustment of calcium excretion occurs in the distal tubule where its reabsorption is promoted by parathyroid hormone. Normally, about 9% of the filtered calcium are reabsorbed distally, with the result that only about 1% of the filtered amount appears in the final urine.

The kidneys thus adjust the excretion of calcium to the quantity absorbed and therefore play an important role in long-term balance. They also regulate the serum calcium concentration and form the active vitamin D metabolite calcitriol.

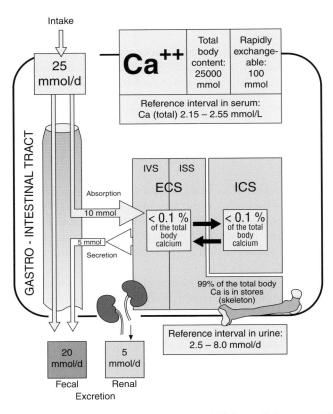

Fig. 22. Calcium balance. ECS: Extracellular space. ICS: Intracellular space. ISS: Interstitial space. IVS: Intravascular space.

Calcium is present in the plasma in three fractions (Fig. 23). About 40% is bound to proteins, predominantly albumin, about 10% is complexed (e.g. to bicarbonate) and about 50% is free ("ionized" calcium). The protein-bound fraction depends on the composition of the plasma proteins and increases with protein concentration. Protein binding is pH-dependent. At a high pH (low proton concentration), more calcium ions are bound and the fraction of ionized Ca^{2+} falls. At low pH the fraction of ionized Ca^{2+} rises.

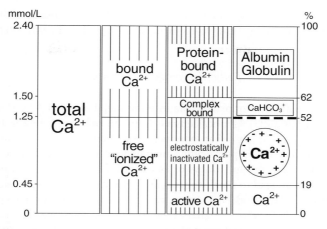

Fig. 23. Calcium fractions in the plasma. Modified from [96].

$$pH < 7.4 \qquad\qquad pH > 7.4$$

$$2[\text{Albumin}^- - H^+] + [Ca^{2+}] \xrightleftharpoons[+2H^+]{-2H^+} [2\,\text{Albumin}^- - Ca^{2+}]$$

The calcium balance is physiologically regulated by parathyroid hormone (PTH, parathyrin) and calcitriol (1a,25-dihydroxycholecalciferol).

The physiological stimulus for parathyroid hormone secretion is hypocalcemia. PTH produces a rise in serum calcium concentration by mobilization of calcium from the bones and by increasing renal reabsorption. Furthermore, it stimulates calcitriol synthesis in the kidneys.

Calcitriol increases calcium absorption from the intestine. Its synthesis in the kidneys from calcidiol (25-hydroxycholecalciferol) is promoted by PTH. In renal failure, calcitriol can only be inadequately produced.

Administration of calcitonin lowers the serum calcium level by inhibiting both osteoclast activity and tubular reabsorption of calcium. It is therefore used therapeutically in hypercalcemia and increased bone turnover (Paget's disease). Calcitonin does not, however, appear to play any essential role in humans (long-term). Neither calcitonin overproduction (in medullary thyroid carcinoma) nor reduced or absent calcitonin secretion (thyroidectomy) leads to any change in the calcium balance.

Pathophysiology and Therapy

Calcium plays an important role in bone metabolism and neuromuscular coupling [93]. Hypocalcemia is found in rickets and tetany. Blood clotting is only impaired at extremely low calcium concentrations not observed in vivo. Hypercalcemia may explain metastatic calcification, e.g. in the eyes, in the kidneys or as chondrocalcinosis, e.g. in the menisci, and can lead to renal stones and nephrocalcinosis (hypercalcemic syndrome). Hypercalcemic crises can be accompanied by tetraplegia and can lead to coma. At higher calcium concentrations the action of the cardiac glycosides is potentiated and at lower concentrations reduced.

Hypocalcemia

The characteristic symptom of hypocalcemia is latent or manifest tetany (Tab. 15). The severity of the symptoms does not always correlate with the calcium concentration; alkalosis, for example, potentiates the symptoms.

Table 15. Symptoms of hypocalcemia

Manifest tetany
Tetanic attacks
Carpopedal spasms, laryngospasm, abdominal cramps, paresthesia

Latent tetany
Trousseau's sign, Chvostek's sign

Central nervous symptoms
Apathy, forgetfulness, grand mal attacks, hallucinations

Cardiovascular symptoms
ECG changes (QT prolongation), arrhythmias, hypotension, heart failure

Chronic trophic disorders
Disorders of bone mineralization, fragile nails, alopecia, dry skin, disorders of dental growth

Basal ganglia calcification

Cataract

Especially patients who develop hypocalcemia rapidly (e.g. after parathyroidectomy) often develop severe symptomatology, while other patients with chronic renal failure can adapt well to hypocalcemia, especially as, in acidosis, the decisive fraction of "ionized" calcium increases. If hypoproteinemia can be ruled out as the cause of hypocalcemia (known as "pseudohypocalcemia"), then the hypocalcemia generally arises from disorders of external balance, e.g. with reduced intestinal calcium uptake or rarely on increased renal calcium loss (Tab. 16).

Hypocalcemia due to disorders of internal balance may be caused by reduced calcium release from the bones or increased precipitation of calcium in the skeleton or soft tissues.

With increased concentrations of complexing agents, e.g. citrate administration during blood transfusion (e.g. liver transplantation), the functionally decisive free fraction is lowered despite "normal" total con-

Table 16. Causes of hypocalcemia

Negative external balance

Reduced intake
• Reduced enteral calcium absorption
 Vitamin D deficiency: alimentary, reduced sunlight exposure, malabsorption
 Insufficient production of active vitamin D metabolites: alcoholism, renal failure, antiepileptic drugs

Increased loss
• Renal loss (loop diuretics)

Disorders of internal balance

Reduced calcium release from bone
• Hypoparathyroidism (idiopathic, after parathyroidectomy, magnesium deficiency)
• Pseudohypoparathyroidism

Increased incorporation of calcium in the bones
• "Hungry bone" syndrome (e.g. after total parathyroidectomy)

Calcium deposition in tissues
• Pancreatitis
• Rhabdomyolysis
• Burns

Hypoalbuminemia/hypoproteinemia

As calcium is 40% protein-bound, the total calcium in hypoalbuminemia is reduced without a fall in ionized calcium and without clinical symptoms. A rule of thumb for correction is that for each deviation of 1 g/L plasma albumin, the total calcium changes by 0.02 mmol/L in the same direction.

centration. This change can, as with pseudohypocalcemia, be ruled out by determining the ionized calcium [45].

Therapy

In tetany: calcium gluconate solution 10% slowly i.v. In the "hungry bone" syndrome after parathyroidectomy: administer 10% calcium gluconate in 500 mL 5% glucose solution over 12–24 h. Treatment is continued as required with calcium salts and vitamin D preparations p.o.

Hypercalcemia

Hypercalcemia exists if the serum calcium level exceeds 2.55 mmol/L. Since the method of determination has been automated, calcium in the serum is very frequently determined routinely. An important consequence has been the discovery of asymptomatic hypercalcemia in up to 0.1% of the total population. 80–90% of these hypercalcemias are caused either by hyperpara-thyroidism or by malignant disease.

A hypercalcemic syndrome is said to be present if the hypercalcemia has led to clinical symptoms (Tab. 17). A hypercalcemic crisis is considered to be present if acute hypercalcemia is accompanied by life-threatening symptoms, with a predominance of cerebral manifestations.

Table 17. Symptoms of hypercalcemia

Cerebral symptoms (hypercalcemic encephalopathy)
Depression, confusion, hallucinations, paranoia, coma

Gastrointestinal symptoms
Anorexia, vomiting, constipation, ulcers, pancreatitis

Cardiovascular symptoms
QT-interval shortening in the ECG, hypertension

Renal symptoms
Poor concentration (polyuria, polydipsia, hypovolemia)
Renal insufficiency (acute and chronic hypercalcemic nephropathy)
Nephrocalcinosis
Nephrolithiasis

Metastatic calcification

Table 18. Causes of hypercalcemia

Positive external balance

Increased intestinal absorption

- Increased vitamin D action
  D-hypervitaminosis
  Vitamin D intoxication
  Endogenous vitamin D hormone overproduction in sarcoidosis
  Granulomatous inflammations
  Idiopathic hypercalcemia in childhood

- Increased calcium supply
  Milk-alkali syndrome
  Calcium-containing medicaments (calcium-containing ion exchangers, phosphate binders)

- Reduced renal excretion
  Thiazides
  Addison's disease
  Verner-Morrison syndrome
  Familial hypocalciuric hypercalcemia

Disorders of internal balance

Increased release of calcium from the bones

- Parathyroid hormone-dependent
  Primary hyperparathyroidism
  Ectopic parathyroid hormone secretion
  "Tertiary" hyperparathyroidism (persistance of a secondary hyperparathyroidism after renal transplantation)
  Lithium therapy (stimulation of the parathyroids?)

- Parathyroid hormone-independent
  Primary bone tumors
  Metastases
  Multiple myeloma and hematological neoplasms
  Increased bone turnover during immobilization
  Hyperthyroidism
  Vitamin A therapy

The serum calcium concentration in these cases is generally over 3.5 mmol/L. There may also be life-threatening central nervous, cardiac and renal disorders. Causes of hypercalcemia are listed in Tab. 18, differential diagnosis is outlined in Tabs. 19 and 20.

Table 19. Tests for differential diagnosis of hypercalcemia

Diagnosis	Phosphate (S)	Intact PTH (S)	Ca^{2+} (U)	Chloride (S)	25(OH)D (S)	1,25(OH)$_2$D (S)
Primary hyperparathyroidism	↓N	↑	↑N	↑	↓N	↑
Familial hypocalciuric hypercalcemia	N	N	↓N	↑N	N	N
Thyrotoxicosis	↑N	↓	↑	↑	N	↓N
Humoral hypercalcemia in malignant disease (PTH-related peptide)	↓N	↓	↑	↓	N	↓
Lymphoma and Hodgkin's disease	↑N	↓	↑	↓	N	↑
Bone metastases	↑N	↓	↑	↓	N	↓
Thiazide therapy	N	↓	↓	↓		
Vitamin D intoxication	↑N	↓	↑	↓	↑	N ↑↓
Milk-alkali syndrome	↑	↓	↓N	↓		
Lithium therapy	↓N	↑	↓	↑		N
Aluminium intoxication	↑	N			N	↓
Sarcoidosis and granulomatous diseases	↑N	↓	↑	↓	N	↑
Immobilization	↑	↓	↑	↓	N	↓

↑ increased ↓ decreased N normal
PTH: Parathyroid hormone
25(OH)D: Calcidiol (25-hydroxy-cholecalciferol)
1,25(OH)$_2$D: Calcitriol (1α,25-dihydroxycholecalciferol)
S: Serum
U: Urine

Treatment of Hypercalcemia

Increase renal calcium excretion: balancing of generally existing fluid deficit and forced diuresis with saline and furosemide. In addition to re-

Table 20. Differential diagnosis of disorders of calcium and phosphate metabolism

	Hypocalcemia	Normocalcemia	Hypercalcemia
Hypophos-phatemia	Hypocalcemia with Hypophosphatemia • Reduced gastro-intestinal Ca/phosphate absorption (vitamin D deficiency) • Increased Ca-phosphate tissue deposits (hungry bone, pancreatitis, rhabdomyolysis)	Normocalcemia with Hypophosphatemia • Reduced phosphate intake and absorption (malnnutrition, malabsorption) • Increased phosphate excretion, renal (phosphate diabetes), gastro-intestinal (diarrhea, vomiting)	Hypercalcemia with Hypophosphatemia • Primary hyperparathyroi-dism • Tumor hypercalcemia • Lithium therapy
Normo-phos-phatemia	Hypocalcemia with Normophosphatemia • Hypoalbuminemia (nephrotic syndrome)	Normocalcemia with Normophosphatemia	Normocalcemia with Normophosphatemia • Thiazide therapy • Familiary hypocalcuric hypercalcemia
Hyper-phos-phatemia	Hypocalcemia with Hyperphosphatemia • Hypoparathyroidism • Renal insufficiency	Normocalcemia with Hyperphosphatemia • Renal insufficiency	Hypercalcemia with Hyperphosphatemia • Bone metastases • Increased intestinal Ca/phosphate absorption (vitamin D intoxication, vitamin D overproduction (sarcoidosis)

The molar product of $S-Ca^{2+}$ times S-phosphate (S-Ca X S-P) is called the "calcium-phosphate product". Normally it is 1.7–5.0 $(mmol/L)^2$. Values below 1.7 $(mmol/L)^2$ may indicate bone demineralization (as in rickets and osteomalacia). A calcium phosphate product greater than 5.00 may give rise to extra-osseous soft tissue calcifications (i.e. in renal insufficiency).

placement of the water loss caused by hypercalcemia (often 5 L and more) an infusion of isotonic saline solution produces, via volume expansion, reduced renal sodium and calcium reabsorption. The efficacy of this forced diuresis is increased with loop diuretics (furosemide). Potassium and magnesium levels must be especially observed. Thiazide diuretics are contraindicated in this case, since they block calcium excretion.

• Reduce enteral absorption of calcium: corticosteroids (e.g. prednisolone 100 mg i.v. per day).

54 Electrolytes in the Serum

- Inhibit bone calcium resorption: calcitonin 400–1000U/day i.v. as infusion, plicamycin (Mithramycin®) 25 pg/kg body weight infused over 6 hours. Repetition of the dose after 24–48 h is possible.
- Bisphosphonates (e.g. clodronate or pamidronate).
- Remove calcium by extracorporeal dialysis with calcium-free dialysate.
- Inhibit prostaglandin synthesis: the involvement of prostaglandins in calcium release in the presence of osteolytic foci is suggested. Indomethacin at a dosage of 75–200 mg/24 h p.o. or acetylsalicylic acid 2–4 g/24 h p.o. are recommended.

1.9 Phosphate

Physiology

Inorganic phosphate is taken to mean the sum of ionized, complexed and protein-bound phosphate that, at a pH of 7.4, is present in the blood predominantly as secondary (hydrogen) phosphate and, to a lesser degree, as primary phosphate (dihydrogen phosphate):

$$H_2PO_4^- \rightleftharpoons H^+ + HPO_4^{2-} \qquad pK = 7.20$$

Phosphate ions play a less important role as a plasma buffer, but a predominant role as an intracellular buffer. Apart from "inorganic" phosphate organically bound phosphoric acid is also present in the plasma.

Of 70 kg body weight about 800 g (= 25800 mmol) are phosphorus (Fig. 24). Inorganic phosphate is predominantly present in the bones and teeth (about 85%), to a lesser extent in muscle cells (14%) and very slightly in extracellular fluid (about 1%).

The daily intake of phosphate is about 40 mmol, mainly with meat, milk products and vegetables. About 30 mmol of this are absorbed, predominantly in the jejunum and promoted by vitamin D. About 25 mmol are excreted renally and about 15 mmol in the feces. The phosphate concentration in the plasma is regulated essentially by the kidneys. The ultrafiltratable fraction is filtered in the glomerulus and reabsorbed in the tubule to about 80%. Parathyroid hormone inhibits renal phosphate reabsorption and increases excretion in the urine; somatotropin, insulin and vitamin D promote tubular phosphate reabsorption.

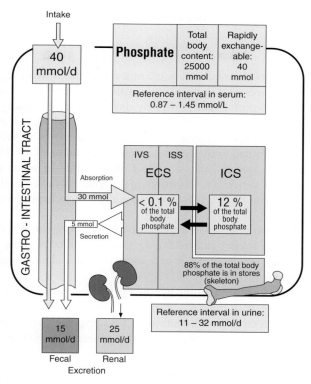

Fig. 24. Phosphate balance. ECS: Extracellular space. ICS: Intracellular space. ISS: Interstitial space. IVS: Intravascular space.

The phosphate level fluctuates in each individual within wide limits and the fluctuations are circadian. Phosphate is involved in almost all intermediate metabolic processes in protein, lipid acid carbohydrate metabolism. The energy-rich phosphate bonds in ATP and creatine phosphate represent energy reserves.

Pathophysiology and Therapy

Hypophosphatemia

If the serum phosphate falls below 0.7 mmol/L, clinically relevant phosphate deficit is present. As phosphate is involved in numerous biological functions, the symptoms of phosphate deficiency are exceedingly numerous and are especially marked in acute hypophosphatemia. In low-grade chronic hypophosphatemia the symptoms are not so variable and they are often limited to the skeletal system and muscles (Tabs. 21, 22).

A negative external balance is most often caused by malnutrition and vitamin D deficiency. Disorders of internal balance, i.e. shift of phosphatase from ECS to ICS, also frequently lead to hypophosphatemia (Tab. 23).

For **treatment** of acute hypophosphatemia potassium hydrogen phosphate or glucose-1-phosphate are given i.v. either at a rate of 0.1 mmol/kg body weight over 6 h or a total of 10 mmol as a long-term infusion over 6 h.

In the treatment of chronic hypophosphatemia (especially in alcoholics) an increase in the daily phosphate intake (p.o. or i.v.) to 30–50 mmol/day is attempted (milk contains 1 g phosphate per liter = 10 mmol phosphate). In contrast to this conservative therapy recommendation

Table 21. Symptoms of acute hypophosphatemia

CNS symptoms
Depression, apathy, coma, confusion, delirium, seizures, motor and sensory neuropathy, dysfunction of cranial nerves, Guillain-Barré syndrome, impaired tissue oxygenation

Hematological symptoms
Hemolysis, thrombocytopenia, reduced chemotaxis

Muscular symptoms
Muscular weakness, myopathy, rhabdomyolysis, respiratory insufficiency, heart failure

Disorders of liver function

Insulin resistance

Table 22. Symptoms in chronic hypophosphatemia

Osteomalacia
Myopathy
Cardiomyopathy

Table 23. Causes of hypophosphatemia

Negative external balance (generally additional to disorders of internal balance)

Reduced intake
• Malnutrition (anorexia, alcoholism, phosphate-free parenteral feeding)
Reduced absorption
• Antacids
• Vitamin D deficiency
• Malabsorption
Increased loss
• Gastrointestinal losses
  Vomiting
  Diarrhea
  Steatorrhea
• Renal loss
  Hyperparathyroidism
  Vitamin D deficiency
  Phosphate diabetes (Fanconi's syndrome and other renal phosphate loss syndromes)

Disorders of internal balance
• Anabolism • Carbohydrate load • Hyperalimentation (especially after malnutrition) • Diabetic ketoacidosis after treatment • Massive cell proliferation (lymphoma, leukoses) • Sepsis • Burns • Alkalosis • "Hungry bone" syndrome after parathyroidectomy

some physicians favor higher dosages of up to 50 mmol phosphate in 6 h, especially in chronic hypophosphatemia with manifest symptoms.

Hyperphosphatemia

Regulation of the phosphate balance occurs mainly via renal excretion, whereby the most important mechanisms are, on the one hand, glomerular filtration of phosphate and, on the other, proximal tubular reabsorption stimulated by parathyroid hormone.

Hyperphosphatemias are therefore almost always attributable to reduced renal phosphate excretion (Tab. 24). This can arise either through

Table 24. Causes of hyperphosphatemia

Positive external balance
Increased intake (with simultaneously reduced phosphate excretion) • Increased oral intake (cow's milk in infants) • Phosphate-containing enemas • Vitamin D administration
Reduced renal excretion • Renal insufficiency (with reduced glomerular phosphate filtration) • Hypoparathyroidism (PTH deficiency) • Pseudohypoparathyroidism • Acromegaly
Disorders of internal balance (release from the intracellular space)
Rhabdomyolysis Cytotoxic drug treatment of malignant diseases

Table 25. Symptoms of hyperphosphatemia

Hypocalcemia Extra-osseous calcification (soft tissue calcification) periarticular, pulmonary, renal, muscular Renal osteopathy

reduced glomerular filtration, as in renal failure, or through increased tubular reabsorption as the result of a lack of parathyroid hormone action.

An increased phosphate intake leads to chronic hyperphosphatemia only in concomitant reduced renal excretion.

Disorders of internal balance in the sense of increased phosphate release from the ICS may lead to acute hyperphosphatemia, but chronic increases in phosphate levels are only observed with additional limitations on renal excretion.

The symptoms of hyperphosphatemia (Tab. 25) are based on the formation of poorly soluble calcium phosphate compounds, which on the other hand results in hypocalcemia and, on the other, can produce extraosseous soft tissue calcification (Tab. 20). Hyperphosphatemia can also promote the development of renal osteopathy via hypocalcemia.

Therapy

Acute hyperphosphatemias are treated by withdrawal of any phosphate-containing substances and administration of calcium, if hypocalcemia exists. In the event of phosphate release from the ICS in the course of cell disintegration through cytotoxic therapy or rhabdomyolysis, adequate fluid therapy is generally sufficient to prevent acute renal failure. Phosphate is dialysable and can be removed by extracorporeal dialysis.

Chronic hyperphosphatemia with severe renal failure must be treated because of the danger of tissue calcification. Dietary phosphate restriction (avoidance of milk and milk products) is generally not sufficient. In order to keep phosphate levels constantly below 2 mmol/L, additional intestinal phosphate binders must be administered. A very effective phosphate binder is, for example, aluminum hydroxide, which should, however, only be used cautiously because of the danger of aluminum intoxication.

Nowadays, the oral administration of calcium salts that form insoluble calcium phosphate compounds in the intestine are preferred. Thus, dialysis patients are given several grams of calcium carbonate or calcium acetate daily to reduce the phosphate level.

2. Electrolytes in the Urine

2.1 Physiological and Pathophysiological Principles

The kidney is the most important organ for regulating the ion composition of the extracellular space [43, 44]. Filtration of primary urine from the plasma takes place via the glomeruli. In the proximal tubule and the pars recta over 70% of the filtered salts and water are reabsorbed, while glucose, amino acids and bicarbonate are almost completely absorbed. In the loop of Henle further reabsorption of water and sodium chloride takes place and, at the transition to the pars convoluta, sodium chloride is again reabsorbed (Fig. 25). The reabsorption of sodium chloride, and indirectly of water, in the distal tubule is regulated by aldosterone.

With reference to the importance of electrolyte determinations in the serum and urine, it should be noted that in most cases the quantity of excreted electrolytes can provide a deeper insight into the pathophysiological course of disorders than the determination of serum concentration.

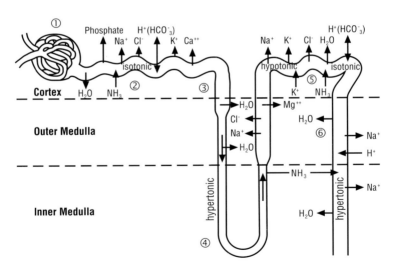

Fig. 25. Absorption and secretion of electrolytes and water in the nephron. 1 Glomerulus, 2 Proximal tubule, 3 Pars recta, 4 Loop of Henle, 5 Pars convoluta, 6 Distal tubule.

Electrolytes in the Urine

The kidney's task is indeed simply to adjust excretion to the requirements and is aimed at constancy of electrolyte concentrations or electrolyte contents in the various spaces. Where disorders are present, the renal excretion is always the first to be adjusted, and only if the kidneys are no longer capable of doing this will the serum concentrations alter.

Measurement of the excretion of the pertinent electrolytes in the urine over a 24 h period generally gives a good indication of the intake in food and the observation of dietary instructions. In parenteral feeding, the necessary intake required for an equilibrated balance can be determined by measurement of the 24 h excretion of certain electrolytes. 24 h urine is generally used to investigate metabolic disorders in nephrolithiasis.

In disorders of electrolyte and acid-base balance important pathophysiological and differential diagnostic indicators can be obtained via urine electrolyte determination [26]. An informative result may often be obtained simply by determining the concentration in the first morning urine or a random specimen. More precise indications are provided by the "urine indices" [92].

Urine Indices

Simply determining electrolyte excretion for the differential diagnosis of disorders of electrolyte balance is, however, often inadequate as it does not take into account the influence of renal function. The urine indices of clearance (C), fractional excretion (FE) and fractional tubular reabsorption JR) have proved to be a more favorable basis for assessing electrolyte balance.

Clearance

The renal clearance (C) provides information about the virtual plasma volume that is cleared per unit of time by renal excretion of a given substance.

The renal excretion is calculated from the product of urine concentration (U) and urine volume (V). If one divides this renally excreted quantity by the plasma concentration (P) of the pertinent substance and the time, then one obtains the plasma volume in which the excreted quantity was previously dissolved and which was cleared of this substance over the corresponding period.

$$C = \frac{U \times V}{P \times t}$$

C   Renal clearance (mL/min)
P   Plasma concentration (mmol/L)
U   Urine concentration (mmol/L)
V   Urine volume (mL)
t   Urine collection time (min)

Generally the clearance (C) is given in mL/min. Urine concentrations (U) and plasma concentrations (P) must be stated in the same concentration units.

The renal clearance of a substance is the plasma volume from which this substance was cleared by the kidney in 1 min. It is dependent on body surface area and can be corrected by multiplication by 1.73/individual body surface area (m^2) in order to permit a comparison with the reference interval [107], which applies to 1.73 m^2. Body surface can be estimated by use of nomograms or according to the following formula:

$$BSA = 0.1672 \, (W \times L)^{0.5}$$

BSA   Body surface area (m^2)
L   Length (m)
W   Weight (kg)

Creatinine Clearance has proved its value in estimating the glomerular filtration rate for over 60 years [85, 88]. In principle, the clearance of all substances that are not protein-bound, not metabolized and neither reabsorbed nor secreted in the tubules is identical to the glomerular filtration rate (GFR). In comparison with other substances (e.g. inulin) creatinine has the advantage of being an endogenous compound whose serum concentration is relatively constant and whose concentration is readily measurable by routine methods.

Although in the event of elevated plasma creatinine concentrations creatinine is not only filtered but also secreted (up to 20%), in medical practice determination of the endogenous creatinine clearance (C_{CR})

Fractional Excretion and Fractional Tubular Reabsorption

The electrolyte excretion in the urine is the net result of the amount filtered by the glomerulus and the amount secreted in the tubules on the one hand and the amount undergoing tubular reabsorption on the other. When calculating the urinary indices the term "fractional excretion" (FE) is used for most electrolytes (Na, Cl, K, Ca, Mg). This gives the amount eliminated in the urine as a fraction of the amount previously filtered by the glomerulus. The respective contributions of tubular secretion and tubular reabsorption cannot be quantified.

For electrolytes not secreted in the tubules (e.g. phosphate) the term Fractional Tubular Reabsorption (FTR) is preferred, which is the difference between the quantity in the glomerular filtrate and the excreted quantity.

The Fractional Excretion (FE) measures the fraction of the primarily glomerularly filtered amount of a substance that appears in the final urine. FE characterizes tubular function. Creatinine clearance (CCR) is used as a measure of the glomerular filtration of electrolytes; the clearance of non-protein bound electrolytes is calculated from their serum concentrations and their excretion in the collected urine (see clearance calculation).

$$\text{FE}(\%) = \frac{C_E}{C_{CR}} \times 100 = \frac{U_E \times P_{CR}}{P_E \times U_{CR}} \times 100$$

C_{CR} Creatinine clearance [mL/min]
C_E Electrolyte clearance [mL/min]
FE   Fractional excretion [%]
P_{CR} Plasma creatinine concentration [mmol/L]
P_E Plasma electrolyte concentration [mmol/L]
U_{CR} Urine creatinine concentration [mmol/L]
U_E Urine electrolyte concentration [mmol/L]

The Tubular Reabsorption (TR) gives the fraction of the amount of a substance filtered by the glomerulus that is reabsorbed and does not appear in the urine. It is expressed by the following mathematical equation:

$$\mathrm{TR}(\%) = \left(1 - \frac{C_E}{C_{CR}}\right) \times 100 = \left(1 - \frac{U_E \times P_{CR}}{P_E \times U_{CR}}\right) \times 100$$

C_{CR}: Creatinine clearance [mU/min]
C_E: Electrolyte clearance [mL/min]
P_{CR}: Plasma creatinine concentration [mmol/L]
P_E: Plasma electrolyte concentration [mmol/L]
TR: Tubular reabsorption [%]
U_{CR}: Urine creatinine concentration [mmol/L]
U_E: Urine electrolyte concentration [mmol/L]

The fractional tubular reabsorption (TR) is the preferred index for characterization of phosphate elimination (TRP). For all other electrolytes the fractional excretion is given instead.

2.2 Sodium Excretion

Sodium is filtered about in the renal glomerulus, so that about 25,000 mmol/day appear in the primary urine. Only about 1% to 3% are excreted in the final urine, i.e. the fractional sodium excretion (FE_{Na}) is 1% (Fig. 26).

Sodium reabsorption in the tubules is, in fact, the main energy-consuming process in the kidneys, and the oxygen consumption in the kidneys correlates directly with Na^+ reabsorption. Any acute damage to the tubules can lead to a reduced sodium reabsorption capacity and thus to an increase in FE_{Na}. In acute renal failure (e.g. acute tubular necrosis), the FE_{Na} is thus generally increased. In hypovolemia and hypoperfusion of the kidneys, on the other hand, sodium is reabsorbed in the kidney to a greater extent, so that less sodium appears in the urine and FE_{Na} is thus generally increased. In hypovolemia and hypoperfusion of the kidneys, on the other hand, sodium is reabsorbed in the kidney to a greater extent, so that less sodium appears in the urine and FE_{Na} is very low (< 1%) (Tab. 26). Sodium determination in the urine is therefore particularly suitable for differential diagnosis of acute renal failure and volume deficiency. Furthermore, determination of sodium in the urine plays an important role in the differential diagnosis of hyponatremia (Fig. 9). Determination of urinary sodium is also used for monitoring after renal

Electrolytes in the Urine

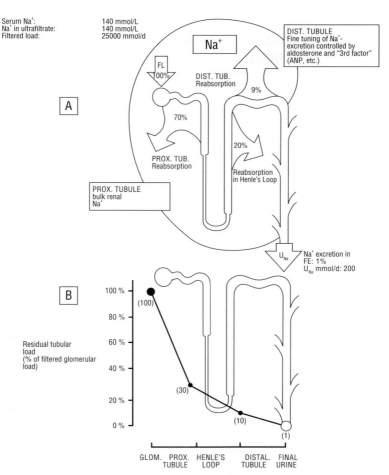

Fig. 26. Intrarenal transport of sodium. A: Fractional reabsorption of sodium in the different nephron segments. B: Fractional amount of sodium remaining in the different nephron segments. ANP: Atrial natriuretic peptide. FE: Fractional excretion (amount excreted as percentage of amount filtered). FL: Filtered load. U_{Na}: sodium excretion in the urine.

transplantation (Tab. 27). A fall of Na⁺ excretion (reduced FE_{Na}) is considered to be an indication of acute rejection reaction.

Table 26. Renal sodium excretion

Renal sodium excretion decreased (= enhanced Na⁺ reabsorption)	Renal sodium excretion increased (= reduced renal Na⁺ reabsorption)

Reduced effective arterial blood volume (secondary hyperaldosteronism)
Total body Na⁺ deficit
  Reduced Na⁺ intake
  Increased extrarenal Na⁺ loss
    Gastrointestinal
    3rd space (ileus, etc.)
    Perspiration
Total body Na⁺ excess
  Heart failure
  Nephrotic syndrome
  Hepatorenal syndrome
  Septicemia
Normal or increased effective arterial blood volume
  Acute rejection episodes
  Mineralocorticoid excess
  Glucocorticoid therapy
	Increased Na⁺ intake
	Renal Na⁺ loss
	  Acute renal failure
	  Na⁺-losing kidney
	  Mineral corticoid deficiency
	  Syndrome of inadequate ADH secretion (SIADH)
	  Diuretics

Table 27. Indications for sodium determination in urine

	$U_{Na} < 20$ [mmol/L] $FE_{Na} < 1\%$	$U_{Na} > 40$ [mmol/L] $FE_{Na} > 3\%$
Acute renal insufficiency	Prerenal failure	Acute tubular necrosis
Volume deficiency	Extrarenal sodium loss	Renal sodium loss
Hyponatremia	Hyponatremia with hypovolemia (severe sodium deficit, see Tab. 2)	Hyponatremia with normovolemia (SIADH)
	Reduced effective volume Hyponatremia with hypervolemia (reduced effective volume: heart failure, liver insufficiency, nephrotic syndrome)	
Renal transplantation	Acute rejection	

U_{Na}:   Sodium concentration in the urine
FE_{Na}:   Fractional sodium excretion in the urine
SIADH: Syndrome of inadequate ADH secretion

Molecular Background

Sodium reabsorption takes place via four different transport systems along the nephron. These are at the same time the sites of action of the different diuretics. Proximal tubular sodium reabsorption is effected mainly by the sodium-hydrogen exchanger (NHE) which exchanges sodium for protons. This is the site where the inhibition of carbonic anhydrase, which catalyses the rapid interconversion of CO_2 and water into carbonic acid, protons, and bicarbonate ions, takes effect.

In the loop of Henle sodium is reabsorbed together with potassium and chloride. The active transporter is the sodium-potassium- chloride co-transporter (NKCC). This is the site of action of the loop diuretics. A genetic defect of this mechanism leads to Bartter's syndrome.

In the distal tubule sodium and chloride are transported together. This is where the thiazide sensitive sodium-chloride co-transporter (TSC) is located and is the site of action of the thiazide diuretics. A genetic disturbance of the TSC causes Gitelman's syndrome.

2.3 Chloride Excretion

Chloride, as the most important counter ion to sodium in the ECS, behaves almost identically to sodium in both the urine and serum. The renal handling of chloride follows the renal transport of sodium. Thus, chloride excretion is normally proportional to sodium excretion.

In the presence of non-absorbable anions in the urine (e.g. penicillin, organic acid anions in metabolic acidosis) sodium is simultaneously excreted as a counter ion and the sodium concentration and sodium indices are therefore higher than the corresponding chloride indices. Differential diagnosis between prerenal failure and renally induced acute renal failure under these circumstances is facilitated by the measurement of chloride excretion, whereby chloride values less than 10 mmol/L indicate prerenal failure and chloride values > 20 mmol/L acute renal failure.

Chloride determination in the urine is clinically important for the differential diagnosis of metabolic alkalosis (Tab. 28). Chloride excretion of less than 10 mmol/L and FE_{Cl} < 0.7 indicates extrarenal chloride loss. Extrarenal chloride losses occur primarily in chronic vomiting and after insertion of gastric probes (gastritic chloride-sensitive alkalosis).

Table 28. Indications for chloride determination in urine

	$U_{Cl^-} < 10$ mmol/L $FE_{Cl^-} < 0.7\%$	$U_{Cl^-} > 20$ mmol/L $FE_{Cl^-} > 1.7\%$
Acute renal insufficiency	Prenal failure	Acute tubular necrosis
Metabolic alkalosis	Chloride sensitive alkalosis	Chloride-resistant alkalosis

U_{Cl^-}: Chloride concentration in the urine
FE_{Cl^-}: Fractional chloride excretion

Other metabolic alkaloses, e.g. mineralocorticoid excess (Conn's disease, Cushing's syndrome, exogenous mineralocorticoid administration), Bartter's syndrome and severe potassium deficiency are not associated with a reduced chloride excretion in the urine. The chloride concentration in urine in such cases is above 10 mmol/L and $FE_{Cl} > 1.7\%$.

2.4 Osmolality

The quantitatively most important osmoles in the urine are the cations sodium, potassium and ammonium with their corresponding anions and urea. Glucose increases the urinary osmolality in pathological states (glucosuria) and, in a similar way, mannitol is effective after therapeutic administration as an osmodiuretic. The urinary osmolality is most often determined by means of freezing point depression. It can be estimated according to the following equation in the absence of exogenously added substances.

$$U_{osm}(\text{mosmol/kg}) = 2 \times ([Na^+] + [K^+] + [NH_4^+]) + [\text{urea}] + [\text{glucose}]$$

Concentrations in mmol/L.

The factor 2 takes into account the concentrations of the pertinent anions.

The amount of osmotically active particles excreted represents the difference between the glomerularly filtered amount and the tubularly reabsorbed amount. Increased excretion can thus arise by increasing the filtered or tubularly secreted amounts or by reducing tubular reabsorption. Within 12 h after complete fluid intake restriction osmolality in urine should exceed 800 to 900 mosmol/kg. Otherwise diabetes insipidus centralis or renalis may be assumed. After application of adiuretin urinary osmolality rises in case of diabetes insipidus centralis, whereas it remains unvariably low in case of renal (tubular) origin.

As with all substances, clearance can also be calculated for the sum of all osmotically active particles.

$$C_{osm} = \frac{U_{osm} \times V}{P_{osm} \times t}$$

C_{osm}: Osmolal clearance [mL/min]
P_{osm}: Osmolality in plasma [mosmol/kg]
t: Urine collection time [min]
U_{osm}: Osmolality in urine [mosmol/kg]
V: Urine volume [mL]

For the assessment of renal function, the use of the urine/serum ratio of osmolality is more informative than the urine osmolality alone in the presence of a changed serum osmolality (especially common in intensive care patients). The osmotic clearance is, however, not an informative parameter for the diagnosis of renal function disorders when considered alone (Fig. 27).

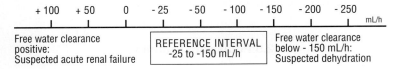

Fig. 27. Free water clearance (C_{H_2O}).

For isotonic urine (when $U_{osm} = P_{osm}$), C_{osm} is equal to the urine volume (V) per minute. In water diuresis U_{osm} is lower than P_{osm}, whereas in antidiuresis U_{osm} exceeds P_{osm}.

The difference between urine volume (V) per minute and osmotic clearance (C_{osm}) is referred to as "free water clearance".

$$C_{H_2O} = \frac{V}{t} - C_{osm} = \frac{V}{t}\left(1 - \frac{U_{osm}}{P_{osm}}\right)$$

C_{H_2O}: Free water clearance [mL/min]
C_{osm}: Osmolal clearance [mL/min]
P_{osm}: Osmolality in plasma [mosmol/kg]

Table 29. Osmolality in urine: Indices and reference intervals

Urine osmolality	50–1600 [mosmol/kg]
• Thirst testing (14 h)	855–1335 [mosmol/kg]
Urine/Plasma osmolality ratio	1.5–3.0
Osmolal clearance	2–4 [mL/min]
Free water clearance (C_{H_2O}):	(– 0.4) to (– 2.5) [mL/min]

t: Urine collection time [min]
U_{osm}: Osmolality in urine [mosmol/kg]
V: Urine volume [mL]

Usually urine is hyperosmolar in comparison with plasma, i.e. C_{osm} is greater than V/t and the free water clearance.

As the kidney loses its ability to concentrate, the free water clearance tends towards zero. At least in septic acute renal failure (and appropriate fluid balance), the loss of concentration ability of the kidneys is an early warning sign, and serial determination of free water clearance represents a useful aid to monitoring.

For indices and reference intervals see Tab. 29.

2.5 Potassium Excretion

Potassium is glomerularly filtered in the kidney and almost completely reabsorbed in the proximal tubule and the loop of Henle (Fig. 28). The amount excreted in the final urine thus arises from the tubular secretion of potassium in the distal and collecting tubules. With normal renal function, excretion during potassium deficiency can be reduced to 10 mmol/day. With potassium excess, by contrast, up to 400 mmol/d may be excreted (Tab. 30).

Molecular Background of Renal Potassium Transport

The physiological regulation of potassium excretion takes place through aldosterone-dependent secretion in the distal part of the cortical collecting duct where it is coupled with sodium reabsorption. The actual potassium channels which allow potassium secretion are called ROMK

Table 30. Renal potassium excretion

Renal potassium excretion decreased	Renal potassium excretion increased

Normo- or hypokalemia
Reduced K$^+$-intake
Hypokalemia
Extrarenal K$^+$-losses
  Gastrointestinal K$^+$-losses
  (vomiting, diarrhea, ileus)
Normo- or hyperkalemia
Blockage of tubular K$^+$-secretion
  Tubular defects (Gordon's syndrome)
  Hypoaldosteronism
  Hyporeninemic hypoaldosteronism
  ACE inhibitor
  Aldosterone antagonists
  K$^+$-Sparing diuretics
  β-Blocking agent
Reduced glomerular filtration
  Renal insufficiency
  (especially with oliguria)
	Hyper- or normokalemia
	Increased K$^+$-intake
	Hypokalemia
	Renal K$^+$-loss
	Polyuria
	K$^+$-losing nephropathy
	Tubular defects (Liddle's syndrome,
	Bartter's syndrome, Pseudo-Bartter)
	Mineralocorticoid excess
	(primary and secondary hyperaldosteronism, Cushing's syndrome, glucocorticoid therapy)
	Diuretics
	(apart from K$^+$-sparing diuretics)

(renal outer medulla-K-channels). Defective ROMKs are the cause of a variant of Bartter's syndrome.

Urine Potassium Concentration

The most important indication for the determination of urine potassium concentration (UK) is the differential diagnosis of hypokalemia (Tab. 30). With renal potassium loss the potassium excretion is relatively high while, in cases of extrarenal potassium loss, the kidney reduces excretion to compensate accordingly.

Extrarenal potassium losses usually lead to a reduction of renal potassium excretion to values below 10 mmol/L. Values greater than 10 mmol/L with existing hypokalemia indicate renal potassium losses (Tab. 31).

Table 31. Indications for potassium determination in urine

Hypokalemia, U$_{K^+}$ < 10 mmol/L	Hypokalemia, U$_{K^+}$ > 30 mmol/L
Reduced potassium intake Extrarenal potassium loss	Renal potassium loss

U$_{K^+}$: Potassium concentration in the urine.

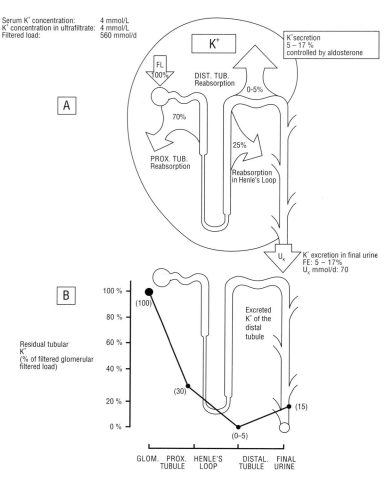

Fig. 28. Intrarenal transport of potassium. A: Fractional reabsorption of potassium in the different nephron segments. B: Fractional delivery of potassium to the different nephron segments. Absorption (and secretion) of potassium in the nephron. FE: fractional excretion in % of filtered potassium; FL: filtered load; U_K: potassium excretion in the urine.

Urinary Sodium/Potassium Ratio (U_{Na}/U_K)

The ratio (U_{Na}/U_K) exceeds physiologically 1. If more potassium than sodium is excreted (ratio < 1) hyperaldosteronism may be suspected.

Potassium Clearance (C_K):

$$C_K = \frac{U_K \times V}{P_K \times U_{CR}}$$

Fractional Potassium Excretion (FE_K):

$$FE_K = \frac{C_K}{C_{CR}} \times 100 = \frac{U_K \times P_{CR}}{P_K \times U_{CR}} \times 100$$

C_{CR}	Creatinine clearance (mL/min)
C_K	Potassium clearance (mL/min)
FE_K	Fractional potassium excretion in the urine (%)
P_{CR}	Plasma creatinine concentration (mmol/L)
P_K	Plasma potassium concentration (mmol/L)
U_{CR}	Urinary creatinine concentration (mmol/L)
U_K	Urinary potassium concentration (mmol/L)

2.6 Anion Gap

The anion gap in the urine is calculated at pH < 6.5 as follows:

> Anion gap in the urine (mmol/L) = [Na$^+$] + [K$^+$] − [Cl$^-$]

All concentrations in mmol/L.

Reference interval of anion gap in urine: 41 ± 9 mmol/L.

As in serum, the excretion of the routinely determined cations (Na$^+$ and K$^+$) in urine exceeds chloride excretion. Bicarbonate, which is always taken into account in calculating the anion gap in serum, can be ignored in urine at a pH of less than 6.5.

The urinary anion gap is a rough estimate for ammonium excretion in urine which may be even negative in case of increased excretion of ammonia (> 80 mmol/d) as an unmeasured cation. It can be a help in the differential diagnosis of hyperchloremic metabolic acidosis. A negative urine anion gap indicates gastrointestinal bicarbonate loss (with maintained renal ammonium excretion) while a positive urine anion gap suggests impaired distal tubular acid excretion [10]. On initiation of metabolic acidosis by ammonium chloride administration, the urinary anion gap is negative in normal individuals. It is also negative in diarrhea, but becomes positive in case of distal renal tubular acidosis.

2.7 Magnesium Excretion

The magnesium concentration in serum does not provide accurate information about the actual magnesium content of the body, especially as the extracellular fraction is only about 1%. Magnesium deficiency may be present even though the plasma magnesium concentration is within the reference interval. Renal magnesium excretion is surprisingly accurate at regulating the total body magnesium content (Fig. 29). In magnesium deficiency excretion in the urine is reduced, with the result that magnesium excretion can be used for diagnosis of magnesium deficiency (especially if the serum magnesium concentration is normal; Tab. 32).

Table 32. Renal magnesium excretion

Renal magnesium excretion decreased	Renal magnesium excretion increased
Magnesium deficiency	Renal magnesium loss
Reduced intake	Mg-losing kidney
Extrarenal loss	(Gitelman's syndrome)
Magnesium shift	Hypercalcemia
"Hungry bone" disease	Hyperthyroidism
	Diuretics
	Aminoglycosides
	Ciclosporine
	Cytotoxic drugs

Electrolytes in the Urine

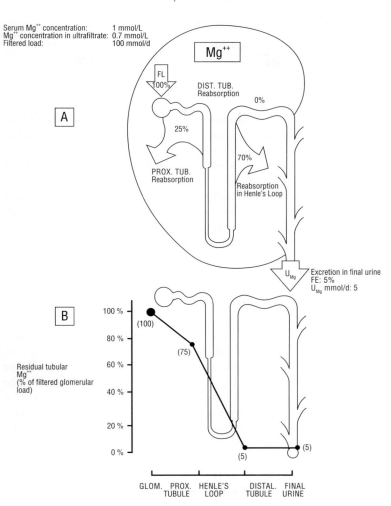

Fig. 29. Intrarenal transport of magnesium. A: Fractional reabsorption of magnesium in the different nephron segments. B: Fractional delivery of magnesium to the different nephron segments. Absorption (and secretion) of magnesium in the nephron. FE: fractional excretion in % of filtered Mg^{2+}; FL: filtered load; U_{Mg}: Mg excretion in the urine.

Table 33. Indications for magnesium determination urine

Hypomagnesemia, $U_{Mg^{2+}}$ < 0.5 mmol/L	Hypomagnesemia, $U_{Mg^{2+}}$ > 1.5 mmol/L
Extrarenal magnesium loss	Renal magnesium loss

$U_{Mg^{2+}}$: Magnesium concentration in the urine.

Furthermore, the measurement of magnesium in the urine is useful for the differential diagnosis between renal or extrarenal causes of magnesium deficit. Increased excretion in the urine with hypomagnesemia indicates renal origin, whereas lowered excretion shows extrarenal origin (Tab. 33).

Since magnesium in the serum, like calcium, is protein-bound and the glomerularly filtered quantity cannot be easily determined, the calculation of urine indices for magnesium is not meaningful.

The determination of magnesium in the urine is necessary in the magnesium loading test (see Fig. 21).

Molecular Background of Renal Magnesium Transport

The renal regulation of Mg excretion has not yet been fully elucidated. Magnesium reabsorption differs from that of all other electrolytes: only 25% is reabsorbed proximally while 70% is reabsorbed in the loop of Henle. In the distal tubule Mg reabsorption is minimal. The reabsorption is only regulated to a small extent via hormonal factors. A greater role is played by intrinsic renal mechanisms, triggered e.g. by a Ca^{2+}/Mg^{2+}-sensing receptor located in the loop of Henle which directly recognizes the changes in the serum Mg concentration.

2.8 Calcium Excretion

Calcium filtered through the glomerulus is reabsorbed to about 95% in the proximal tubule, a process which is promoted by parathyroid hormone and inhibited by sodium ions. With a normal diet, the calcium excretion in the urine is 2.5–8.0 mmol/d (Fig. 30).

Hypercalciuria can develop if, as a result of hypercalcemia (Tab. 18), the quantity of glomerularly filtered calcium rises (Tab. 34). Hypercal-

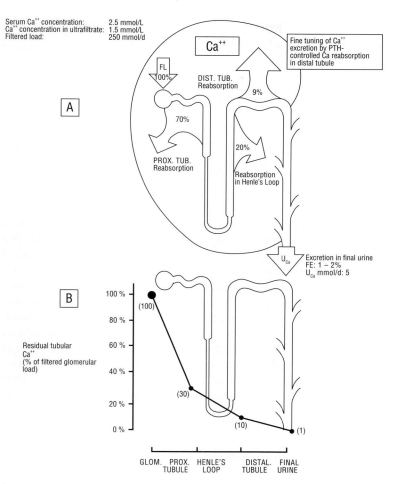

Fig. 30. Intrarenal transport of calcium. A: Fractional reabsorption of Ca^{++} in the different nephron segments. B: Fractional delivery of Ca^{++} to the different nephron segments. Absorption (and secretion) of Ca^{++} in the nephron. FE: fractional excretion in % of filtered Ca^{++}; FL: filtered load; U_{Ca}: Ca excretion in the urine.

cemia may be caused by disorders of internal balance (see calcium in the serum) and also by a positive external balance (e.g. increased intestinal absorption). It is seen in renal tubular acidosis, hyperthyroidism and

Table 34. Renal calcium excretion

Renal calcium excretion decreased	Renal calcium excretion increased
Hypocalcemia	**Hypercalcemia**
Hypoparathyroidism	Renal tubular acidosis
Normo- or hypercalcemia	Tumor hypercalcemia
Familial hypocalciuric hypercalcemia	Osteolytic process
Milk-alkali syndrome	Sarcoidosis
Thiazide therapy	Hyperparathyroidism
	Thyrotoxicosis
	Vitamin D intoxication
	Cushing's syndrome
	Normocalcemia
	Loop diuretics

Cushing's syndrome. The urine calcium determination plays a subordinate role in the differential diagnosis of hyper- and hypocalcemia. The diagnostic value is limited by the fact that, in one and the same clinical picture, calcium excretion is simultaneously promoted and hindered by various adverse pathophysiological influences. Primary hyperparathyroidism, for example, leads to hypercalcemia by release of calcium from the bones with an increase in calcium excretion, but, on the other hand, parathyroid hormone favors calcium reabsorption in the distal tubule. On the whole, the calcium excretion-promoting factors predominate in primary hyperparathyroidism, so that hypercalciuria generally results.

The main significance of calcium determination in urine lies in the clarification of nephrolithiasis. About one third of all patients with calcium stones have hypercalciuria.

Molecular Background of Renal Calcium Transport

Tubular reabsorption of calcium takes place passively in the proximal tubule and in the loop of Henle, usually by a paracellular route and always together with sodium. Dissociation between sodium and calcium transport does not occur until in the distal tubule where the physiological fine-tuning of calcium excretion takes place under the influence of parathyroid hormone, calcitonin and vitamin D. The transport system consists inter alia of dihydropyridine-sensitive Ca channels. Calcium-sensing receptors (CaSR) may be able to control renal calcium reabsorption directly in response to changes in the calcium concentration in the extracellular space.

2.9 Phosphate Excretion

The kidneys play an important role in the regulation of phosphate homeostasis. After glomerular filtration more than 80% is normally reabsorbed (tubular phosphate reabsorption, TPR). Tubular secretion does not appear to be of any importance in humans.

Renal phosphate excretion therefore depends on two mechanisms, the glomerular filtration of phosphate and its tubular reabsorption (Fig. 31, Tab. 35). Dietary fluctuations are offset primarily by changes in the amount filtered. Changes in phosphate reabsorption are caused predominantly by hormonal factors. Parathyroid hormone reduces tubular reabsorption, whereas vitamin D promotes it.

Determination of phosphate excretion in the urine alone is of little differential diagnostic value as it depends on so many factors (e.g. phosphate intake, bone metabolism, renal function). For this reason, the following parameters are used to assess phosphate elimination:

- Phosphate clearance (C_p)
- Fractional tubular phosphate reabsorption (TRP [%])
- Threshold value for phosphate excretion (TmP/GFR)

Table 35. Renal phosphate excretion

Renal phosphate excretion decreased	Renal phosphate excretion increased
Reduced glomerular filtration	**Increased phosphate intake**
Acute renal insufficiency	
Chronic renal insufficiency	**Tubular phosphate loss**
Increased tubular reabsorption	Phosphate diabetes
Reduced phosphate intake	Primary and secondary tubular defects
Increased phosphate needs	Renal tubular acidosis
Growth, pregnancy, lactation, acromegaly	Hyperparathyroidism (primary and secondary)
Hypoparathyroidism	

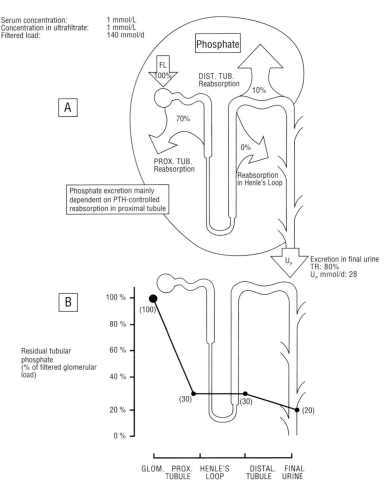

Fig. 31. Intrarenal transport of phosphate. A: Fractional reabsorption of phosphate in the different nephron segments. B: Fractional delivery of phosphate to the different nephron segments. Absorption (and secretion) of phosphate in the nephron. FE: fractional excretion in % of filtered phosphate; FL: filtered load; U_P = phosphate excretion in the urine.

Electrolytes in the Urine

Molecular Background of Renal Phosphate Transport

Regulation of phosphate excretion takes place mainly in the proximal tubule. The molecular basis of this phosphate transport is the sodium-phosphate cotransporter (Na-P_i-(= inorganic phosphate) co-transporter Type II) in the brush border of the proximal tubule. The activity of this Na-P_i co-transporter is reduced in various hypophosphatemia/hyperphosphaturia syndromes. The gene for NaP_i II has been located on chromosome 5. A regulating PEX gene on the X chromosome is held responsible for the frequently occurring X-linked hypophosphatemias.

Phosphate Clearance

$$C_P = \frac{U_P \times V}{P_P \times t}$$

C_P: Phosphate clearance [mL/min]
P_P: Plasma phosphate concentration [mmol/L]
t: Urine collection time [min]
U_P: Urine phosphate concentration [mmol/L]
V: Urine volume [mL]

The phosphate clearance is dependent on the alimentary phosphate (and NaCl) intake, the excretory function of the kidney (glomerular filtration rate) and hormonal factors influencing tubular reabsorption.

Fractional Tubular Phosphate Reabsorption (TPR%)

$$TRP(\%) = \left(1 - \frac{C_P}{C_{CR}}\right) \times 100$$

C_{CR}: Creatinine clearance [mL/min]
C_P: Phosphate clearance [mL/min]

By transformation one obtains:

$$\text{TRP}(\%) = \left(1 - \frac{U_P \times P_{CR}}{P_P \times U_{CR}}\right) \times 100$$

P_{CR}: Plasma creatinine concentration [mmol/L]
P_P: Plasma phosphate concentration [mmol/L]
TRP (%): Fractional tubular reabsorption of phosphate
U_{CR}: Urine creatinine concentration [mmol/L]
U_P: Urine phosphate concentration [mmol/L]

The urine volumes are canceled out in the equation, with the result that TPR (%) can be determined from spontaneous urine.

The TPR has the advantage over phosphate clearance in that it takes into account renal function (CCR). However, as the TPR is also dependent on phosphate intake a reduction in TPR is seen with increased intake.

TmP/GFR (Threshold Value for Phosphate Excretion)

The maximum tubular phosphate reabsorption is identical to the theoretical renal threshold for phosphate excretion. TmP/GFR is considered the best parameter for determining tubular disorders of phosphate reabsorption. Renal excretion starts only above a given plasma concentration, namely when the amount filtered exceeds the maximum reabsorption rate.

Unfortunately, this renal threshold concentration is imprecise as not all nephrons exhibit exactly the same transport maximum, and therefore, if only small amounts of phosphate are excreted, the filtration curve for the kidney as a whole exhibits a feature known as "splay". At phosphate excretions above this splay the curve is linear and one can assume that, with increased phosphaturia, or with a lower-grade tubular phosphate reabsorption (TPR < 80%), the reabsorption capacity of all nephrons is exhausted and that the tubular maximum is achieved. The maximum tubular transport rate can readily be calculated from the difference between the amount filtered per unit time and the amount excreted per unit time.

$$\text{TmP} = \text{Filtered amount/t} - \text{excreted amount/t}$$
$$\text{TmP} = P_p \times \text{GFR} - U_p \times V \times t^{-1}$$

GFR: Glomerular filtration rate [L/min]
P_p: Plasma phosphate concentration [mmol/L]
t: Urine collection time [min]
TmP: Tubular maximum of phosphate reabsorption [mmol/min]
U_p: Urine phosphate concentration [mmol/L]
V: Urine volume [L]

The renal threshold phosphate concentration is achieved when the amount of phosphate filtered is equal to the tubular maximum of phosphate reabsorption.

Threshold phosphate concentration × GFR = TmP
Threshold phosphate concentration = TmP/GFR

$$\text{TmP/GFR} = P_p - \frac{U_P \cdot P_{CR}}{U_{CR}}$$

GFR: Glomerular filtration rate [L/min]
P_{CR}: Plasma creatinine concentration [mmol/L]
P_p: Plasma phosphate concentration [mmol/L]
TmP: Tubular maximum phosphate reabsorption [mmol/min]
U_p: Urine creatinine concentration [mmol/L]
U_p: Urine phosphate concentration [mmol/L]

For indices and reference intervals see Tab. 36.

If phosphate excretion is high (TPR < 80%), the splay of the filtration curve is exceeded, with the result that the phosphate threshold can be calculated.

Table 36. Phosphate in urine: indices and reference intervals

Phosphate excretion	11–32 mmol/d
Fractional tubular phosphate reabsorption (TPR)	82–90%
Phosphate clearance (C_p)	5.4–16.2 mL/min
TmP/GFR (threshold phosphate concentration)	0.8–1.4 mmol/L

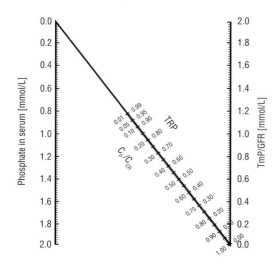

Fig. 32. Nomogram from Walton and Bijvoet for the estimation of renal threshold phosphate concentration [122]. C_{CR}: creatinine clearance (mL/min), C_P: phosphate clearance (mL/min), GFR: glomerular filtration rate (L/min), TmP/GFR: threshold phosphate concentration (mmol/L), TmP: tubular maximum phosphate reabsorption (mmol/min), TPR: fractional tubular phosphate reabsorption (%).

If the phosphate excretion is low (TPR > 80%), then the nomogram of Walton and Bijvoet is used for estimating threshold phosphate concentration (Fig. 32):

Serum or plasma phosphate concentrations and pertinent TPR or CP/CCR values are marked and joined with a ruler so as to intersect the TmP/GFR axis. The intersection gives the threshold phosphate concentration of the patient.

The phosphate threshold is reduced in tubular syndromes with phosphate loss and also in primary and secondary hyperparathyroidism.

3. Acid-Base Balance and Blood Gases

3.1 Physiology of Acid-Base Balance

The acid and base equivalents occurring during metabolism are produced mainly by the liver (Fig. 33). Altogether the complete combustion of carbohydrates and fats produces about 16 mol/day of volatile carbon dioxide (CO_2) which is exhaled via the lungs. The incomplete combustion of these macronutrients and the metabolization of acid amino acids (glutamic acid, aspartic acid) and of sulfur- and phosphorus containing proteins produces acid equivalents while the breakdown of basic amino

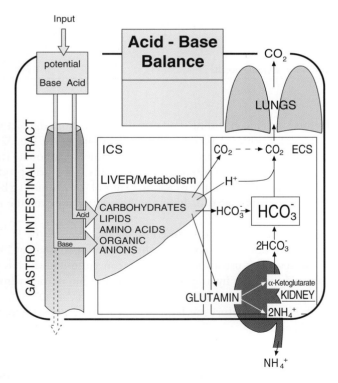

Fig. 33. Acid-base balance. ICS: intracellular space, ECS: extracellular space.

Table 37. Net proton balance in acid-base metabolism as a result of breakdown of carbohydrates, fats, proteins and metabolites

Substance	Biochemical Reaction and Reaction Products	Net Proton Balance (mmol/day)
Carbohydrates, fats	complete combustion to CO_2 (16,000 mmol/d)	±0
	incomplete combustion	+30
Amino acids	neutral amino acids with non-ionized groups (-OH, -SH, -CO-NH$_2$; serine, threonine, alanine, valine)	±0
	neutral sulfide containing amino acids (cystine, methionine)	+70
	basic amino acids (lysine, arginine, histidine)	+130
	acid amino acids (asparaginic acid, glutamic acid)	−100
Metabolites: citrate, acetate		−60
Fecal elimination of citrate, acetate		0–(−60)

acids (lysine, arginine, histidine) and organic anions from the diet (e.g. citrate, lactate) produces base equivalents (bicarbonate, see Tab. 37). It is estimated that on an average diet about 160 mmol bicarbonate (total alkali gain) and about 230 mmol acid equivalents (total acid production) occur daily. The difference of about 70 mmol (or 1 mmol/kg body weight) must be eliminated daily by the kidneys as non-volatile acid [25, 35].

In spite of marked changes caused by the metabolism, the H$^+$ ion/proton concentration in the blood remains remarkably constant at 35–43 nmol/L. However, it is usually not the proton concentration which is given but the pH (potentia hydrogenii; negative logarithm of the hydrogen ion activity). This is kept constant at between 7.37 and 7.45 (37 °C).

The chief compensatory mechanisms which are responsible for this constant pH are:

- Reversible binding of the protons by buffers.
- Respiratory regulation by adjusting the exhalation of carbon dioxide.
- Renal regulation by bicarbonate regeneration ("renal acid excretion").

This is also the order in which the mechanisms take effect (Tab. 38): buffering immediately; respiratory counterregulation within minutes, renal regulation within days.

Table 38. Buffer mechanisms and the order in which they take effect

extracellular buffering	effective immediately
pulmonary CO_2 elimination	effective within minutes
intracellular buffering	within 2–4 hours
renal acid excretion	after hours to days

In an ionogram the cations and anions in the plasma are shown side by side and can be readily compared (Fig. 34).

The blood plasma is composed (inter alia) of the bases NaOH, KOH, Ca(OH)$_2$, Mg(OH)$_2$ (in the respectively shown amounts) and the acids H_3PO_4, H_2SO_4, HCl and proteins. The bicarbonate ion (HCO_3^-) is in a steady state with the undissociated acid H_2CO_3 and with CO_2. The regulation of this steady state is catalyzed by carbonic anhydrase.

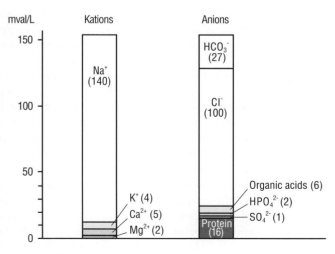

Fig. 34. Ionogram of the blood plasma (see also Fig. 2).

$$H_2CO_3 \leftrightarrow H^+ + HCO_3^- \leftrightarrow CO_2 + H_2O$$

The HCO_3^-/H_2CO_3 system is the most important buffering system in the plasma. At the pH of blood (7.4) the ratio of HCO_3^- to H_2CO_3 is 20:1. According to the Henderson-Hasselbalch equation the pH is dependent not on the absolute amounts but on the ratio of the concentrations of undissociated acid/anion (pK). The pK of carbon dioxide (first step) is 6.1; in the case of a non-volatile acid the best buffering action would be expected at this pK. However, as CO_2 is volatile and an enzymatically catalyzed steady state between CO_2 and H_2CO_3 is always established, the effective H_2CO_3 concentration is dependent chiefly on the carbon dioxide tension (pCO_2). The great surplus of HCO_3^- in the buffering system of the blood now proves to be of advantage as H^+ ions are captured and bound to OH^- ions as water while CO_2 escapes in gas form. In the other direction the buffering capacity is practically inexhaustible; used up H_2CO_3 is regenerated immediately as CO_2 is always available in excess.

As HCO_3^- is equivalent to the amount of a strong acid whose H^+ ions can be buffered according to the equation $H^+ + HCO_3^- = H_2CO_3$, this quantity is referred to – not quite correctly – as alkali reserve. It is normally about 25 mmol/L. A state in which the alkali reserve is reduced is referred to as compensated acidosis as long as the pH is still normal; if the pH falls we speak of decompensated acidosis. The opposite state, in which the HCO_3^- is increased, is referred to as alkalosis.

Two further buffering systems are involved in maintaining a constant blood pH. Firstly the phosphate system $HPO_4^{2-}/H_2PO_4^-$ is a good buffer in this pH range, corresponding to a pK of 7 for the second dissociation step of phosphoric acid. Secondly the hemoglobin of the red blood cells acts as a buffer; oxygenation increases the acid strength, H^+ is dissociated off and thus more CO_2 is eliminated via the lungs; if oxygenation is low, less CO_2 is released via the lungs.

On account of the paramount importance of the bicarbonate buffer system particularly for the extracellular space, where the routine determinations of the acid-base balance are performed, the most important parameters of the acid base balance are the components of the bicarbonate buffer system $CO_2 + H_2O \leftrightarrow H_2CO_3 \leftrightarrow H^+ + HCO_3^-$.

For reference intervals see annex.

Acid-Base Balance and Blood Gases 89

Table 39. Additional parameters of the acid-base balance

- **Base excess (BE)**
  The BE is the difference between the actual and normal buffer bases of a sample.
  Reference interval: −2–(+3)
- **Buffer bases (BB)**
  Sum of all buffer anions in whole blood which are active at a medically relevant pH. The value is thus dependent on the hemoglobin and phosphate concentrations.
  Reference interval 44–48 mmol/L.
- **Standard bicarbonate**
  Bicarbonate concentration in plasma or whole blood which has been equilibrated in vitro with a pCO_2 of 40 mm Hg and a pO_2 of > 100 mm Hg at 37 °C.
  Reference interval: 21–26 mmol/L.

The three parameters pH, pCO_2 and HCO_3^- are usually sufficient for a differential diagnosis of the acid-base balance. Further parameters can be helpful in individual cases (Tab. 39).

The pH of the 'transcellular fluids' differs in some case considerably from that of blood. Gastric juice is very acid (pH 1.5), the contents of the small intestine are alkaline at around pH 8, urine is usually slightly acid at around pH 5.

3.2 Pathophysiology of Acid-Base Balance: Introduction

Disturbances of the acid-base balance are divided into acidoses and alkaloses depending on the pH and into respiratory and metabolic disorders depending on the cause. In the majority of cases the three components of the bicarbonate buffer system (pH, pCO_2, HCO_3^-) are sufficient for classification (Tab. 40).

Table 40. Changes in pH, bicarbonate (actual) and pCO_2 in acidosis and alkalosis

	Metabolic	Respiratory
Acidosis	pH ↓	pH ↓
	HCO_3^- ↓	HCO_3^- ↑
	pCO_2 ↓	pCO_2 ↑
Alkalosis	pH ↑	pH ↑
	HCO_3^- ↑	HCO_3^- ↓
	pCO_2 ↑	pCO_2 ↓

Respiratory disturbances are caused by a disturbance of CO_2 exhalation. If exhalation of CO_2 is reduced there is increased formation of carbonic acid and an increase in the proton and bicarbonate concentrations in the blood. A respiratory acidosis is thus characterized by an increase in pCO_2 and HCO_3^- and a decrease in the pH. In reverse, increased exhalation of CO_2 leads to a fall in pCO_2 and HCO_3^- accompanied by an increase in pH. These are the characteristic features of respiratory alkalosis.

Metabolic disorders are caused primarily by alterations of the proton or bicarbonate concentrations in the extracellular space (ECS). Increased production of protons, with consequent consumption of the bicarbonate buffer, or primary bicarbonate losses lead to metabolic acidosis. The characteristics of the metabolic acidoses are therefore a reduced bicarbonate concentration together with a decrease in pH. With normal respiratory regulation there is increased exhalation of CO_2 which causes the pCO_2 to fall.

Metabolic alkaloses are caused by acid losses – the bicarbonate concentration and the pH increase (the H^+ ion concentration falls), followed by respiratory regulation in the form of hypoventilation and thus an increase in the pCO_2.

In most cases the direction of the pH shift indicates whether the primary disorder is acidosis or alkalosis (Fig. 35).

In metabolic disorders there is a primary disturbance of the pH or bicarbonate concentration with secondary respiratory compensation via CO_2 exhalation. In patients with healthy lung function there is a strict correlation between pH or bicarbonate and pCO_2. This makes it possible to describe the compensatory changes.

In uncomplicated metabolic acidosis or metabolic alkalosis the pCO_2 values shown in Tab. 41 are measured.

In respiratory disorders there is primarily a disturbance of CO_2 ventilation, i.e. a primary change in the pCO_2. This is followed by secondary renal compensation by adjustment of bicarbonate production. It takes several days for these compensatory mechanisms to become effective. Depending on the duration of the respiratory (pCO_2) changes, various bicarbonate concentrations can therefore be expected (see Tab. 42).

Acid-Base Balance and Blood Gases

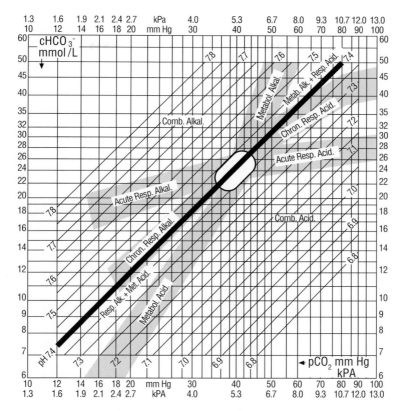

Fig. 35. Nomogram for diagnosis of acid-base disorders taking into account the degree of compensation.

pCO_2 is plotted logarithmically on the abscissa, the bicarbonate concentration logarithmically on the ordinate. If there is a disturbance with a normal degree of compensation the intersection of the measurements is within the respective field. If the intersection lies outside, this can be due to one of the following reasons:
– The disturbance has only been present for a short time and compensation has not yet had time to occur
– The function of the compensating organ, e.g. the lungs in the case of metabolic disturbances, the kidneys in the case of respiratory disturbances, is impaired. Adapted from Müller-Plathe [86].

Table 41. Extent of compensation in metabolic acidosis and alkalosis

Metabolic acidosis	Expected pCO_2 [mm Hg] = $1.5 \times HCO_3^-$ (mmol/L) + 8	
	Rule of thumb: if the pH is above 7 the expected pCO_2 corresponds to the 2 places after the decimal point of the pH (pCO_2 = 7... mm Hg).	
	Example: Metabolic acidosis pH: 7.25 (measured) pCO_2: 25 mm Hg (expected as a result of compensation)	
Metabolic alkalosis	Expected pCO_2 [mm Hg] = $0.95 \times HCO_3^-$ (in mmol/L) +15	
	Rule of thumb: the expected pCO_2 corresponds to the 2 places after the decimal point of the pH	
	Example: Metabolic alkalosis pH: 7.55 (measured) pCO_2: 55 mm Hg (expected as a result of compensation)	

Table 42. Extent of compensation in respiratory acidosis and alkalosis

Acute respiratory acidosis	Expected HCO_3^- -increase	+1 mmol/L per 10 mm Hg pCO_2-increase[1]
Chronic respiratory acidosis	Expected HCO_3^- -increase	+3 mmol/L per 10 mm Hg pCO_2-increase[1]
Acute respiratory alkalosis	Expected HCO_3^- -decrease	−2 mmol/L per 10 mm Hg pCO_2 -decrease[1]
Chronic respiratory alkalosis	Expected HCO_3^- -decrease	−5 mmol/L per 10 mm Hg pCO_2 -decrease[1]

[1] Based on [HCO_3^-] of 25 mmol/L and pCO_2 of 40 mm Hg.

3.3 Metabolic Acidosis

Acidoses are divided into respiratory and metabolic acidoses. Respiratory acidoses are caused by insufficient elimination of CO_2 via the lungs. Metabolic acidoses are caused by increased endogenous production or exogenous intake of acids or loss of alkaline secretions. Determination of the anion gap makes it possible to distinguish between the different causes of metabolic acidosis. In every form of acidosis compensatory mechanisms come into action to keep the pH at a constant level. If there is metabolic acidosis with an accumulation of non-volatile acids, a simultaneous decrease in the pCO_2 can be expected. In reverse, if there is prolonged respiratory acidosis, we can expect to see an increase in the bicarbonate concentration as a result of renal compensation.

Metabolic acidoses are caused by accumulation of non-volatile acids or by primary bicarbonate losses. Metabolic disorders are characterized by shifts in pH and bicarbonate which move in the same direction (e.g. a low pH accompanied by a decrease in the bicarbonate concentration = metabolic acidosis).

Metabolic acidosis: pH ↓
HCO_3^- ↓
pCO_2 ↓

The extent of the pCO_2 decrease indicates whether or not sufficient respiratory counterregulation (compensation) has taken place.

Symptoms of Metabolic Acidosis

Table 43. Symptoms of acute and chronic metabolic acidosis

Symptoms of acute and chronic metabolic acidosis	
Cerebrovascular symptoms:	Increased intracranial pressure, headaches
Cardiovascular symptoms:	Tachy-/bradycardia, arrhythmia, negative inotropism, hypotension
Pulmonary symptoms:	Kussmaul breathing, pulmonary hypertension, pulmonary edema
Gastrointestinal symptoms:	Intestinal atony
Symptoms of chronic metabolic acidosis	
Renal symptoms:	Hyperkaliuria, hyperkalemia, hypercalciuria, nephrocalcinosis, nephrolithiasis (including spontaneous steinstrasse)
Skeletal symptoms:	Short stature in children, osteopenia
Hormonal disturbances:	Reduced secretion of parathyroid hormone, erythropoietin Reduced activity of renal 1α-hydroxylase (see vitamin D) Increased secretion: catecholamines, glucocorticoids, renin, aldosterone
Metabolism:	Increased proteolysis, insulin resistance, increased lipolysis

Causes of Metabolic Acidosis

The main characteristic of the metabolic acidoses is a reduced plasma bicarbonate concentration following increased bicarbonate consumption. This can be the result either of an increased exogenous (ingestion) or endogenous (metabolism) acid load (= addition acidosis) or of increased bicarbonate losses (subtraction acidosis, see Tab. 44). Bicarbonate losses can be of intestinal or renal origin. The main job of the kidneys in the context of maintaining acid-base homeostasis is the recovery/generation of bicarbonate.

Table 44. Causes of metabolic acidosis

Increased acid load (addition acidosis)	
Ingestion of exogenous acids:	Hydrochloric acid
	Ammonium chloride
	Arginine hydrochloride
	Lysine hydrochloride
Ingestion of exogenous acid precursors:	Salicylic acid
	Ethyl alcohol (ethanol)
	Methyl alcohol (methanol)
	Ethylene glycol
	Paraldehyde
Loss of bicarbonate (subtraction acidosis)	
Renal causes:	Uremia
	Proximal tubular acidosis
	Distal tubular acidosis
Gastrointestinal causes:	Diarrhea
	Fistulas
	Loss of intestinal secretions (pancreas, bile)
	Calcium chloride
	Magnesium sulfate
	Colestyramine
Ureterosigmoidostomy	

Differential Diagnosis of Metabolic Acidosis

For differential diagnosis it is useful to subdivide the metabolic acidoses into:

- Metabolic acidoses with increased anion gap
- Hyperchloremic metabolic acidoses

To preserve electroneutrality a reduction in serum bicarbonate (the main characteristic of metabolic acidosis) must be accompanied by an increase in other anions. Therefore the decrease in bicarbonate must be followed either by an increase in chloride (hyperchloremic acidosis) or by an increase in the concentration of the not normally measured anions – indicated by an increased anion gap (high anion gap acidosis).

High Anion Gap Metabolic Acidosis

High anion gap metabolic acidosis results from the ingestion or increased endogenous production of organic acids.

The names of the conditions which lead to high anion gap acidosis form the acronym KUSMALE where K stands for ketoacidosis (diabetic, alcoholic, starvation), U for uremia, S for salicylate poisoning, M for methanol poisoning, A for alcohol poisoning, L for lactic acidosis and E for ethylene glycol poisoning (Tab. 45).

Pathophysiologically speaking, the high anion gap metabolic acidoses are addition acidoses since there is increased production (endogenous) or ingestion (exogenous) of organic acids. Uremic acidosis forms an exception being an acidosis caused primarily by a bicarbonate deficit. The increased anion gap is caused by diminished glomerular filtration of anions.

Table 45. High anion gap metabolic acidoses

Ketoacidoses	Diabetes, alcohol, starvation
Uremia	End-stage renal failure
Salicylate poisoning	
Methyl alcohol poisoning	
Ethyl alcohol poisoning	
Lactic acidosis	Hypoxic (circulatory failure, ischemia), liver failure, septicemia, pancreatitis
Ethylene glycol poisoning	

Hyperchloremic Metabolic Acidosis

The hyperchloremic metabolic acidoses are caused by ingestion of hydrochloric acid (HCl) or chlorides of amino acids, e.g. lysine, methionine, by primary bicarbonate deficiency as a result of intestinal or renal bicarbonate losses or reduced bicarbonate regeneration by the kidneys (Tab. 46).

Table 46. Causes of hyperchloremic metabolic acidosis

Exogenous ingestion of acids	Mineral acids or their salts, e.g. HCl, ammonium chloride, lysine chloride, arginine chloride, methionine etc.
Toluene	Glue sniffers
Bicarbonate loss	Renal: moderately severe renal failure, proximal renal tubular acidosis (RTA Type 2), distal renal tubular acidosis (RTA Type 1, 3, 4) Gastrointestinal: diarrhea, intestinal loss of secretions (pancreatic, biliary, intestinal) Calcium chloride, magnesium sulfate, colestyramine Ureterosigmoidostomy

The hyperchloremic metabolic acidoses are for the most part addition acidoses (ingestion of mineral acids) and subtraction acidoses (bicarbonate losses). Toluene poisoning in glue sniffers is a special case in that we are talking here of poisoning by an organic acid (hippuric acid is produced from toluene). However, the hippurate anion is eliminated so quickly that it does not accumulate in the serum (normal anion gap).

3.4 Metabolic Alkaloses

The main characteristics of the metabolic alkaloses are:

Metabolic alkalosis: pH ↑
HCO_3^- ↑
pCO_2 ↑

The actual pCO_2 indicates whether sufficient respiratory counterregulation has taken place (i.e. whether or not a compensated acid-base disturbance is present). After respiratory compensation a $pCO_2 = 0.95 \times HCO_3^- + 15$ (see Tab. 41) is measured.

Causes of Metabolic Alkalosis

Metabolic alkaloses are caused either by increase alkali ingestion (addition alkaloses) or by acid losses (subtraction alkaloses). Losses of gastric fluid lead to alkalosis on account of the hydrochloric acid it contains. Renal acid losses increase the bicarbonate concentration (Tab. 47).

Table 47. Generation of metabolic alkaloses

Alkali loading	Oral and parenteral alkali loading (milk alkali syndrome, ingestion or administration of sodium bicarbonate, potass97ium carbonate, lactate, citrate, acetate)
Acid losses	Gastric acid losses Intestinal acid losses (chloridorrhea) Renal acid losses

Metabolic alkalosis can only be sustained if, in spite of the alkalosis, the kidneys reabsorb bicarbonate or if bicarbonate generation is not suppressed.

In hypovolemia the increased amount of bicarbonate cannot be eliminated. Volume depletion leads to maintenance of metabolic alkalosis. Administration of sodium chloride to correct the volume deficit corrects the alkalosis.

The most common cause of chloride-responsive metabolic alkalosis is gastric alkalosis in which, in addition to the HCl, large amounts of fluid are lost and the extracellular fluid (ECF) becomes depleted. Administration of physiological saline completely corrects this alkalosis.

Some metabolic alkaloses cannot be corrected by volume expansion. In these cases the alkalosis is maintained by other mechanisms, particularly by hypokalemia and mineralocorticoid excess, which stimulate renal bicarbonate generation. In spite of increased serum bicarbonate levels bicarbonate generation is not suppressed and the metabolic alkalosis is sustained (Tab. 48).

Differential diagnosis of metabolic alkalosis see Tab. 49.

98   Acid-Base Balance and Blood Gases

Table 48. Maintenance of metabolic alkalosis

Chloride responsive metabolic alkalosis	Through volume depletion and persisting renal bicarbonate reabsorption
Chloride resistant metabolic alkalosis	Metabolic alkalosis which cannot be corrected by volume repletion, usually with hypokalemia

Table 49. Differential diagnosis of metabolic alkalosis

Chloride responsive metabolic alkalosis	**Alkali gain**
	Oral and parenteral alkali loading
	  Milk-alkali syndrome
	  Sodium carbonate, calcium carbonate
	  Lactate, citrate, acetate
	  Posthypercapnic alkalosis
	Acid losses
	Gastric acid losses (gastric alkalosis)
	  Chronic vomiting
	  Secret vomiting (anorexia nervosa)
	  Stomach drainage
	Intestinal acid losses (chloridorrhea)
	Renal acid losses (bicarbonate generation)
	  Hypovolemia due to diuretic therapy

Chloride resistant metabolic alkalosis (hypokalemic metabolic alkalosis)	**Renal potassium losses** Primary hyperaldosteronism (adenoma, bilateral hyperplasia, carcinoma) Glucocorticoid-remediable aldosteronism Secondary hyperaldosteronism (Bartter, pseudo-Bartter, hepatic cirrhosis) Renal artery stenosis, Liddle's syndrome Apparent mineralocorticoid excess and liquorice abuse

3.5 Respiratory Acidoses

Respiratory acid base disorders are caused by a primary disturbance of CO_2 elimination. Hypoventilation with increased CO_2 retention leads to a reduction in the pH and an increase in bicarbonate (pH and bicarbonate move in opposite directions = respiratory disorder). Respiratory acidosis is characterized by:

$$\boxed{\begin{array}{l} \text{pH} \downarrow \\ HCO_3^- \uparrow \\ pCO_2 \uparrow \end{array}}$$

In contrast to the numerous causes of metabolic acidosis respiratory acidosis is therefore always caused by hypoventilation. In chronic respiratory acidoses the compensatory mechanisms located in the respiratory centers of the brain are usually damaged. In addition to pulmonary diseases, diseases of the nervous system and the musculature are the most common causes of hypoventilation (Tab. 50).

Table 50. Causes of respiratory acidoses

Mechanical disturbances of the respiratory apparatus	Multiple rib fractures, elevation of the diaphragm (e.g. after upper abdominal surgery with subsequent gastrointestinal atony, ileus, peritonitis), tracheal stenosis after thyroidectomy, tracheomalacia, hematothorax and hang-over from muscle relaxants
Damage of the lung parenchyma	Postoperative or posttraumatic lung collapse, aspiration, pulmonary edema, pneumonia, emphysema
Disturbances of the respiratory centre	Head injury, drug effects (overdose of hypnotics, narcotics)
Neuromuscular diseases	Guillain-Barré syndrome, poliomyelitis, tetanus, muscular dystrophy, polymyositis

3.6 Respiratory Alkaloses

Respiratory alkalosis is characterized by primary hyperventilation with a lowering of pCO_2 and serum bicarbonate, the pH is raised:

$$\begin{array}{c} pH \uparrow \\ HCO_3^- \downarrow \\ pCO_2 \downarrow \end{array}$$

Hypoxia leads to stimulation of the respiratory centre. This can lead to increased alveolar ventilation and thus to respiratory alkalosis. In a number of lung diseases which lead to hypoxia respiratory alkalosis can be found on account of the respiratory compensation. This is particularly typical of pulmonary fibrosis, pulmonary edema, pulmonary embolism and acute asthma attacks. Psychogenic hyperventilation is very common. Diseases of the CNS can also cause hyperventilation. Acute cerebral disorders such as craniocerebral trauma, encephalitis, occasionally also ischemic stroke are particularly relevant here. Stimulation of the respiratory center is a known side-effect of salicylates. Sympathicomimetics and theophylline preparations also cause respiratory alkalosis. Pyretic states can also trigger marked hyperventilation, particularly in the phase of the acute temperature increase. Lastly, respiratory alkalosis can occasionally be the result of inadequate mechanical ventilation. Hyperventilation is used therapeutically for the treatment of cerebral edema (Tab. 51).

Table 51. Causes of respiratory alkalosis

Hypoxemia and primary lung diseases	Reduced pO_2 Severe anemia Acute and chronic lung diseases Pulmonary embolism, pulmonary edema	
Primary stimulation of the respiratory centre	Functional hyperventilation	Anxiety, tension, delirium
	Organic	Stroke, tumor, infections, neurological diseases, trauma
	Medication	Salicylates, catecholamines
	Hormones	Hyperthyroidism, pregnancy
	Various	Pyrexia, gram-negative septicemia, heat, liver failure
Mechanical hyperventilation		

3.7 Treatment

An acid-base disorder is not a disease in its own right. There is always an underlying disease. The primary therapeutic goal is therefore not the acute, rapid correction of the acid-base balance but the treatment of the underlying disease.

In the case of respiratory disturbances a respiratory acidosis is treated by elimination of the hypoventilation or by mechanical ventilation. In metabolic disorders elimination of the cause (insulin deficiency in diabetic ketoacidosis, volume depletion in gastric alkalosis) is also the primary aim. If the underlying condition cannot be remedied – which is often the case in chronic conditions – the acid-base status must be addressed directly, e.g. in renal acidosis.

Acidosis

Causal treatment for correction of acidosis consists in

- Insulin treatment for diabetic ketoacidosis
- Treatment of shock/hypoxia in lactic acidosis caused by shock/hypoxia
- Correction of an obstructive ventilatory disorder in respiratory acidosis
- Mechanical ventilation in hypoventilation as a result of neurological diseases
- Toxin elimination in intoxications

Symptomatic treatment is usually only indicated in metabolic acidoses. Respiratory acidoses cannot be treated symptomatically by administration of alkaline substances but only by rectification of the underlying respiratory failure.

Symptomatic treatment of metabolic acidosis is necessary if a causal treatment does not take effect quickly enough. Symptomatic treatment is performed by administration of alkaline substances. Here the type of substance used, the mode of administration and the dose are important.

Various citrate salts are available for oral administration, namely preparations containing calcium citrate, potassium citrate or sodium citrate. If possible, sodium bicarbonate should not used for oral therapy as CO_2 is formed on reaction with gastric acid and the resulting gas may produce gastrointestinal discomfort. For patients with advanced renal failure calcium citrate is the most suitable preparation. Large amounts of potassium citrate should not of course be given. In patients with concomitant heart failure large doses of sodium citrate should be avoided. Sodium bicarbonate is usually used for intravenous treatment. Administration of tris buffer is obsolete.

Chronic metabolic acidoses without life-threatening symptoms are corrected by oral administration. Intravenous correction of an acidosis within several hours is only necessary in acute acidoses and/or in the event of severe central nervous or cardiovascular symptoms or life-threatening hyperkalemia as a result of the acidosis.

The amount of bicarbonate required for correction of an acidosis can be calculated using the following rule of thumb: $0.3 \times$ base excess (mmol/L) × body weight (kg). In longstanding acidoses, in which increased alkali reserves have also been lost from the skeleton, this dose is not sufficient.

Alkaloses

A mild alkalosis with a pH of up to 7.5 often does not require treatment as there are no clinical symptoms or clinically relevant electrolyte shifts. This applies to alkaloses caused by diuretic or steroid administration, for example.

Marked alkalosis caused by permanent vomiting or loss of gastric secretions is an exception. In these chloride-responsive alkaloses the administration of volume and saline, e.g. in the form of isotonic NaCl solution, with or without additional KCl depending on the serum K^+ concentration, is sufficient to correct the alkalosis. Additional administration of acid is not usually necessary.

In respiratory alkaloses alveolar ventilation must be reduced. This can be achieved by calming the patient or by administration of a sedative, e.g. diazepam, if necessary.

3.8 Oxygen: Physiology

The organism requires a continuous supply of oxygen in order to be able to survive and to prevent permanent damage to its organs, e.g. the brain.

The factors actors influencing oxygen delivery to the tissues are:

- Oxygen content and pressure of the inhaled air
- Oxygen uptake by the lungs
- Oxygen transport in the bloodstream
- Tissue perfusion

Atmospheric Oxygen

The partial pressure of oxygen in the air is 19.9 kPa (149 mm Hg) at sea level, 11.6 kPa (87 mm Hg) at an altitude of 4000 m and 6.1 kPa (46 mm

Hg) at 8000 m. It is only possible to stay at an altitude of 8000 m for a limited time and only after appropriate acclimatization.

Pulmonary Gas Exchange

The processes in pulmonary gas exchange include:

- Ventilation
- Diffusion
- Perfusion
- Distribution

Ventilation

The major determinant of the gas exchange is the functional alveolar ventilation (V(A)). V(A) is that part of the minute volume (amount of gas expired per minute) which takes part in the gas exchange. If ventilation is unimpaired V(A) for oxygen is 250–300 mL/min.

Diffusion

Diffusion describes the process by which the gas molecules enter the lung capillaries down the pressure gradient. The alveolar pO_2 is 13.3 kPa (100 mm Hg), that in the blood flowing into the pulmonary arteries 5.3 kPa (40 mm Hg). During the contact time of about 0.3 seconds oxygen molecules rapidly pass into the blood so that the blood leaving the lungs also has a pO_2 of 13.3 kPa. For normal, effective diffusion the following conditions must be fulfilled:

Surface area available for gas diffusion: 50–100 m²

Thickness of respiratory membrane: (alveolar epithelium, interstitium, capillary endothelium, plasma, red cell membrane) 1 µm

Pressure difference: 8 kPa

Perfusion

For rapid transport of the O_2 molecules the lung tissue must be sufficiently well perfused with blood. It is also important that alveolar ventilation (normal: 4.5 mL/min) and perfusion (normal: 5.0 mL/min) match. The ventilation-perfusion ratio is normally 0.9 (± 0.1).

Distribution

If there is a good local match between alveolar ventilation and perfusion the distribution is optimal. However, even physiologically there are small segments of lung tissue in which ventilation is reduced while perfusion is unimpaired so that there are always mild ventilatory distribution disturbances. In the reverse case there would be a circulatory distribution disturbance.

Both ventilatory distribution disturbances and venoarterial shunts occur physiologically and lead to a pO_2 of 10.7–13.3 kPa (80–100 mm Hg) in the pulmonary vein.

Oxygen Transport

The solubility of oxygen in blood is poor, far worse than that of CO_2. The solubility coefficient for O_2 in blood at 37 °C is 0.0105 mmol × L^{-1} × kPa^{-1} or 0.014 mmol × L^{-1} × $Torr^{-1}$, which is equivalent to 0.235 mL × L^{-1} × kPa^{-1} or 0.0313 mL × L^{-1} × $Torr^{-1}$ (CO_2: 0.230 mmol × L^{-1} × kPa^{-1} in plasma, 37 °C).

Thus at 37 °C and a pO_2 of 12 kPa (90 mm Hg) 1 liter of blood contains only 2.8 mL dissolved oxygen, which is 1.4% of the total O_2 content of the blood. Instead, almost the entire oxygen transported in the blood is hemoglobin bound: approx. 200 mL/L blood (8.9 mmol/L) at a normal hemoglobin concentration. The oxygen capacity, i.e. the maximum volume of oxygen that can be bound by a particular amount of hemoglobin, can be estimated as follows:

$$O_2 = C_{Hb} \times 1.34$$

O_2: oxygen (mL)
C_{Hb}: hemoglobin concentration (g/100 mL)
1.34: Hüfner's constant. This empirically determined factor gives the amount of O_2 (in mL) that is bound by 1 g hemoglobin.

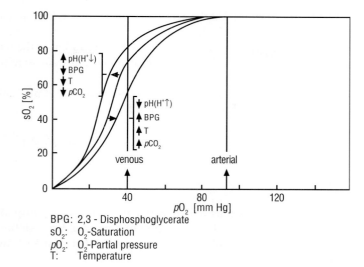

Fig. 36. Oxygen binding curve of hemoglobin.

The binding of oxygen to the hemoglobin in the red blood cells as a function of the oxygen partial pressure is described by the oxygen binding curve (Fig. 36). This curve has a sigmoid shape. This means that even when there is a marked reduction in the pO_2 (e.g. 8 kPa (60 mm Hg)) almost complete oxygen saturation of the hemoglobin can still be achieved (0.91). On the other hand in oxygen-poor tissue the majority of the oxygen bound to the hemoglobin can be released.

The half-saturation pressure of hemoglobin is often given instead of the oxygen dissociation curve. The p_{50} is the oxygen partial pressure present in a hemoglobin solution with an oxygen saturation of 50%. For the saturation range 40% to 90% p_{50} can be calculated using Hill's equation:

$$\log p_{50} = \log pO_2 - \frac{\log[sO_2/(1-sO_2)]}{n_{Hill}} n$$

pO_2: O_2-partial pressure
sO_2: O_2-saturation
n_{Hill}:   2.7

The actual p_{50} is determined at the actual pH and $pCO2$ and the actual temperature. The actual p_{50} is thus dependent on pH, pCO_2 and temperature as well as the 2,3-diphosphoglycerate concentration in the red cells and the proportion of abnormal hemoglobins and dyshemoglobins.

Standard p_{50} ($p_{50,std}$) refers to the half-saturation pressure at pH 7.40, pCO_2 5.33 kPa and 37 °C. It is calculated as follows:

$$\log p_{50,std} = \log p_{50} + \phi_H (7.40 - pH) + \phi_c (\log 5.33 - \log pCO_2)$$

ϕ_H: proton Bohr factor
ϕ_c: carbamate Bohr factor

In contrast to p_{50}, $p_{50,std}$ is thus only dependent on the 2,3-diphosphoglycerate concentration and the proportion of abnormal hemoglobins and dyshemoglobins. In 'healthy' subjects $p_{50,std}$ values between 3.12 and 3.89 kPa have been measured.

An increase in the half-saturation pressure reflects a shift to the right of the oxygen dissociation curve, a decrease a shift to the left. Oxygen binding, and thus the oxygen dissociation curve and the half-saturation pressure, are influenced by the following factors:

- Temperature
- Acid-base status (pH, pCO_2)
- 2,3-Diphosphoglycerate concentration (red blood cells)
- Dyshemoglobins
- Abnormal hemoglobins

Temperature

The dissociation of oxygen from hemoglobin increases with the temperature:

$$\Delta \log pO_2 = 0.024 \times (T - 37)$$

T: actual body temperature (°C)
$\Delta \log pO_2$: $\log pO_{2(50)37\,°C} - \log pO_{2(50)T}$

That is, hypothermia makes oxygen delivery to the tissues more difficult, pyrexia facilitates oxygen delivery.

Acid-Base Status

The dissociation of oxygen from hemoglobin increases with increasing pCO_2 and increasing H^+ activity, i.e. with decreasing pH (classical Bohr effect). In addition, it should be noted that oxygenated hemoglobin (Hb-O_2) is a stronger acid than deoxygenated hemoglobin (HHb) (Haldane or Bohr effect). This fact has the following consequences:

Tissue: Oxygen is passed from the blood to the tissues, HbO_2 is converted to HHb. HHb as weaker acid binds protons so that increased HCO_3^- is formed from H_2CO_3, approx. 0.35 mol per mol of O_2 delivered to the tissues.

Lungs: HHb is oxygenated to HbO_2 which, as stronger acid, gives off protons which react with HCO_3^- to form H_2CO_3 which dissociates further into $CO_2 + H_2O$. CO_2 is exhaled via the alveolae.

The following also plays a role in CO_2 transport from the tissues into the lungs:

HHb binds CO_2 as carbamate to its terminal amino group. During oxygenation in the lungs the carbonate dissociates again and the resulting CO_2 can be exhaled. This effect alone is believed to permit the transport of about half of the CO_2 produced in the tissues.

2,3-Diphosphoglycerate

The synthesis of 2,3-diphosphoglycerate (DPG) in the red blood cells is facilitated by an intraerythrocytic decrease in H^+ activity (increase in pH). The small increase in the pH occurring when HbO_2 is converted to HHb is already sufficient to stimulate DPG production. However, the changes in the DPG concentration do not occur immediately but with a latency of between a few hours and a few days. This mechanism counteracts the increase in the oxygen affinity of hemoglobin occurring with increasing pH.

DPG synthesis is also subject to end-product inhibition. As DPG is bound more strongly to HHb than to HbO_2 the production of DPG is also facilitated by this effect when there is an increase in deoxygenated hemoglobin (HHb).

A reduced serum phosphate concentration leads to a decrease in the DPG concentration, an increased serum phosphate concentration to an increase.

Dyshemoglobins

CO-hemoglobin

The ratio of CO-Hb to HbO_2 depends on the respective partial pressures:

$$\frac{CO-Hb}{HbO_2} = \frac{pCO}{pCO_2} \times M$$

At pH 7.35 M is 300, i.e. the affinity of hemoglobin for CO is 300 times higher than the affinity for oxygen. If there is acidosis or alkalosis M decreases. The blockade of hemoglobin for O_2 transport by the formation of CO-Hb is aggravated by the fact that a concomitant shift to the left of the oxygen dissociation curve leads to a decrease in p_{50}:

CO-Hb: 2%, p_{50} 3.51 kPa (26.3 mm Hg)
CO-Hb: 25%, p_{50} 2.40 kPa (18.0 mm Hg)
CO-Hb: 50%; p_{50} 1.55 kPa (11.6 mm Hg)

An increase in CO-Hb of 1% leads to a decrease in p_{50} of approximately 0.033 kPa (2.5 mm Hg).

This means that, in addition, the oxygen is more firmly bound to the remaining HbO_2, thus increasing the hypoxemia. If, for example, 50% of the hemoglobin is present as CO-Hb, the disturbance of oxygen delivery to the tissues is greater than the disturbance that would be caused by a 50% decrease in the hemoglobin concentration.

CO is a major product of incomplete combustion of carbon and carbon-containing compounds. The installation of catalytic converters has significantly reduced the CO content of car fumes.

Methemoglobin

Methemoglobin (hemiglobin) is unable to transport oxygen. As in the case of CO-hemoglobin, the oxygen binding in the remaining oxyhemoglobin is more firmly bound and thus oxygen delivery to the tissues is additionally worsened. However, the effect is only about half as pronounced as in the case of CO-hemoglobin. The methemoglobin constantly produced by the metabolism under physiological conditions is reduced to hemoglobin by the intraerythrocytic methemoglobin reductase so that in healthy subjects only a small methemoglobin fraction is measured (0.8%, smokers 2.7%, based on total hemoglobin in each case).

An increase in the methemoglobin fraction is only rarely due to an inborn deficiency of methemoglobin reductase. It is usually caused by extraneous substances: aniline derivatives (paints), nitro compounds (e.g. nitrobenzene/paint solvents), benzene, nitrite (meat preservation) and drugs (phenazone, acetanilide, phenacetin, sulfonamides).

Abnormal Hemoglobins

Abnormal hemoglobins are characterized by an abnormal amino acid sequence of the globin chain. At present more than 300 abnormal hemoglobins are known, some having increased and some decreased oxygen affinity. An increased oxygen affinity is often accompanied by a compensatory erythrocytosis while a decreased affinity is often accompanied by anemia. Measurement of an extremely abnormal $p_{50,std}$ should always raise the suspicion of abnormal hemoglobins. $p_{50,std}$ is 1.6 kPa (12 mm Hg) in patients with Hb Rainier and 9.3 kPa (70 mm Hg) in patients with Hb Kansas. In addition, the shape of the oxygen dissociation curve is also often altered. Hill constants of between 1.0 and 2.9 have been found in the presence of abnormal hemoglobins instead of 2.7.

In beta-thalassemia the increase in oxygen affinity is commensurate with the extent to which beta chains are replaced by gamma chains, resulting in formation of HbF in place of HbA_o.

Oxygen Delivery to the Tissues

Under resting, fasting conditions 250–300 mL O_2/min are taken up by the human organism. The tissues with the highest oxygen requirement per 100 g tissue (approx. 10 mL/min) are cardiac muscle, cerebral cortex and renal cortex.

3.9 Oxygen: Pathophysiology

Pulmonary Gas Exchange

Ventilation

Alveolar hypoventilation is accompanied by a fall in the arterial pO_2 and an increase in pCO_2. As both pO_2 and pCO_2 are pathological we speak here of combined hypoxemic/hypercapnic respiratory failure. Ventilatory disorders are divided into

- obstructive disorders, e.g. in bronchial asthma, chronic obstructive bronchitis, obstructive pulmonary emphysema, cystic fibrosis
- restrictive disorders, e.g. due to thorax deformities, muscle diseases, pleural effusion, pleural fibrosis, interstitial lung diseases, pulmonary fibrosis, pulmonary resection.

Diffusion

In disturbances of diffusion the arterial pO_2 is lowered. Since, under the same conditions, diffusion of 20 times more CO_2 than O_2 is possible, disturbances of diffusion do not lead to an increase in pCO_2. We thus have a hypoxemic, non-hypercapnic insufficiency of lung function.

Disturbances of diffusion which are caused by prolongation of the diffusion pathway occur in pulmonary fibrosis, sarcoidosis or pulmonary congestion. However, usually the primary cause of the reduced diffusion capacity is a reduction in the surface area available for gas exchange as in emphysema and pulmonary sclerosis, for example.

Perfusion

In perfusion disorders significant volumes of venous blood bypass the lung, e.g. in right-to-left shunts or malformations of the heart or the large vessels.

Distribution

Disturbances of ventilatory distribution occur in obstructive ventilatory disorders and postoperatively. Circulatory distribution disturbances are seen after pulmonary embolism, for example.

Oxygen Transport

Under pathophysiological conditions there are usually no medically significant changes in oxygen binding to hemoglobin in the lungs. On account of the sigmoid shape of the oxygen dissociation curve, even marked acidosis or marked increases in the 2,3-diphosphoglycerate (DPG) concentration only lead to a minimal reduction in the oxygen saturation of hemoglobin.

On the other hand, pathophysiologically relevant changes in oxygen release often occur in:

- Hypoxemia
- Anemia
- Acidosis – alkalosis
- Massive transfusion

Hypoxemia

Chronic hypoxemia is typically accompanied by a reduced O_2 affinity or increased p_{50} on account of increased DPG concentrations. The proportion of HHb is increased compared with HbO_2 which leads to an intraerythrocytic increase in pH. Both result in an increased DPG concentration: HHb binds DPG more strongly than HbO_2 and thus reduces the end product inhibition of DPG synthesis. The pH increase activates the enzymes of DPG synthesis and inhibits the enzyme of DPG degradation. Chronic hypoxemia is found in heart failure, cyanotic heart defects and cardiopulmonary diseases, for example.

Anemia

A reduced O_2 affinity is also found in numerous forms of anemia. It is caused by an increased DPG concentration which is due to an increased fraction of HHb in comparison with HbO_2. At reduced Hb concentrations oxygen extraction is enhanced and the HHb fraction increases.

Acidosis – Alkalosis

An increase in the hydrogen ion activity decreases the O_2 affinity of hemoglobin, a decrease in hydrogen ion activity increases O_2 affinity (Bohr effect). Prolonged pH shifts trigger counter-regulation via a change in the DPG concentration. A decrease in the intraerythrocytic pH inhibits DPG production and increases the oxygen affinity, an increase in the pH has the opposite effect, so that the Bohr effect is lessened but not completely eliminated.

Longstanding pH shifts should only be corrected gradually. If the pH is corrected rapidly the Bohr effect is lost and only the counterregulatory change in the DPG concentration is effective. After rapid normalization of a low pH, for example, the DPG concentration is initially still low and there is consequently a marked increase in oxygen affinity which considerably impedes oxygen delivery to the tissues.

Massive Transfusion

When blood is stored the DPG concentration falls. In massive transfusion it must therefore be assumed that there is a decrease in the oxygen affinity of the hemoglobin and that improvement in O_2 delivery is not commensurate with the amount of blood administered.

Oxygen Delivery

The oxygen delivery to the tissues can be disturbed for the following reasons:

Arterial Hypoxia

This is usually caused by insufficient pulmonary arterialization of the blood. However, the oxygen supply is only seriously threatened if pO_2 falls below 4.0 kPa (30 mm Hg).

Anemic Hypoxia

This form of hypoxia occurs if insufficient hemoglobin can be made available for oxygenation, e.g. in severe anemia but also in CO poisoning and methemoglobinemia.

Affinity Hypoxia

The increased binding of the oxygen to hemoglobin disturbs the delivery of oxygen to the tissues, e.g. in alkalosis, in massive transfusion of DPG-poor red cells, in the presence of certain abnormal hemoglobins. CO poisoning and methemoglobinemia are aggravated by the associated high oxygen affinity.

Ischemic Hypoxia

This occurs if there is underperfusion of the tissues which can be caused locally by vascular occlusion or a disturbance of the microcirculation or alternatively by a reduced cardiac output, e.g. in heart failure.

4. Preanalysis

4.1 Sodium

Patient

Fasting blood is not necessary for sodium determination. The plasma sodium concentration is independent of the subject's body position when collecting the sample. There is little [127], if any [15], circadian rhythm for sodium concentration in the serum.

Specimen

Venous and capillary blood show no significant concentration differences and, similarly, neither do serum or heparinized plasma [even when using sodium heparinate in the recommended concentrations (15 IU/mL blood)]. Despite the low sodium concentration in the erythrocytes (about 16 mmol/L [25]), slight hemolysis does not lead to any lowered values of clinical relevance because of the large absolute reference interval. On mixing 3 volumes of hemolyzed erythrocytes with 97 volumes of serum the sodium concentration decreases, for example, from 140 mmol/L to 136.3 mmol/L (−2.7%). The hemoglobin content of the serum in this case is about 1000 mg/L. The sodium concentration does not change if the blood is stored for 24 h at room temperature. Early separation of the serum and storage at +4 °C or for storage over several months −20 °C with protection from evaporation is, however, recommended.

4.2 Chloride

Patient

The chloride concentration in the serum at any given moment is scarcely influenced by food intake or by the patient's body position when the sample is collected [91]. A weak circadian rhythm is present [127].

Specimen

Venous or capillary blood [53], and serum or heparinized plasma can be used equally well. Hemolysis leads to only a very slight error. On mixing 3 volumes of hemolyzed erythrocytes (chloride: 50 mmol/L) with 97 volumes of serum (chloride: 100 mmol/L) the chloride concentration falls to 98.5 mmol/L (−1.5%). The concentration does not alter if the serum is stored for 7 days in a refrigerator. During storage of blood, a chloride shift occurs, i.e. a shift of chloride into the erythrocytes in exchange for hydrogen carbonate. For this reason, serum should be separated as soon as possible from the blood clot.

4.3 Osmolality

Patient

See "Sodium" and "Chloride".

Specimen

The serum should be separated from the blood clot as soon as possible after the sample collection in order to avoid electrolyte shifts. As well as serum, plasma obtained by addition of heparin may be used. If the samples are protected from CO_2 loss and evaporation, the osmolality of serum (plasma) does not change during storage for 24 h at 4 °C [14].

4.4 Potassium

Patient

The time of food intake is of secondary importance to the potassium concentration in the blood, so that the subject does not have to be in the fasted state when the sample is collected. Blood samples taken in the sitting position give about 4% higher potassium concentrations than in the supine position. There is said to be a circadian rhythm which leads to a 10% lower concentration at about 5 p.m. compared to 8 a.m. Repeated closure of the fist by the subject during stasis when taking a blood specimen leads to a rise in the potassium concentration.

Specimen

The potassium concentrations in venous and capillary blood are not significantly different. During analysis of serum higher concentrations are found than when using heparinized plasma while, at the same time, the intra- and interindividual scatter increases. The clotting process is associated with a varying marked retraction of the blood clot in which potassium is released from the erythrocytes and platelets. For this reason, plasma is to be preferred as the specimen for potassium determinations. Because of the high concentration differences between intra- and extracellular space and the (absolute) small reference interval, even a slight (in vitro) hemolysis can lead to a massive error in the potassium concentration. Three volumes of hemolyzed erythrocytes (potassium: 92 mmol/L) with 97 volumes of serum (potassium: 4.0 mmol/L) mixed together produce a rise in serum concentration to 6.64 mmol/L (+ 66%). Consequently, macroscopically visible hemolytic samples are unsuitable for potassium determination. Plasma should be obtained immediately after collecting the sample by careful centrifugation (1200 g, 20 °C) and should be separated at once from the blood constituents. It can then be stored until analysis for several weeks if protected from evaporation without any resulting change in potassium concentration.

The same applies to serum after completion of clotting. On storage of blood in the refrigerator potassium passes, even without release of hemoglobin, out of the erythrocytes (no hemolysis) in accordance with the concentration gradient as the corresponding ion pumps are not functioning, and this leads to falsely high values.

During in vivo hemolysis, thrombocytosis and chronic myelosis, potassium is released from the cells to produce "pseudohyperkalemia". In these cases plasma should be used instead of serum in order to minimize the artifacts.

4.5 Magnesium

Patient

The magnesium concentration in the serum at a given moment is not influenced by food intake. As a result of protein binding, it is to be expected that, like calcium, blood sample collection in a seated subject will lead to higher concentrations in plasma than collection from supine patients. However, as the protein-bound fraction is lower compared to calcium, the effect will be expressed to a lesser extent. A circadian rhythm of serum concentration has not been reliably demonstrated [15].

Specimen

The magnesium concentrations in capillary and pertinent venous blood do not differ significantly in statistical terms. Magnesium concentration in the serum is up to 0.1 mmol/L higher than in heparinized plasma. Hemolysis leads to falsely high magnesium concentrations. If 3 volumes of hemolyzed erythrocytes (magnesium: 2.72 mmol/L) and 97 volumes of serum (magnesium: 1.00 mmol/L) are mixed, then the magnesium concentration in the serum rises to 1.05 mmol/L (+ 5%). For this reason the plasma or serum should be separated as soon as possible. The magnesium concentration in serum does not alter during a 7-day storage period in a refrigerator.

4.6 Total Calcium

Patient

The calcium concentration in the serum at a given moment is not essentially influenced by food intake.

If the blood is collected from the cubital vein of seated subjects, then the calcium concentrations in the serum are 2% higher than in samples collected from supine patients, as a result of a rise in protein-bound fraction. The calcium concentration is subject to a circadian rhythm with an amplitude of 0.1 mmol/L [81].

Specimen

In the serum from venous blood identical or 5% lower [14] concentrations are measured than in capillary blood. During storage of serum in the refrigerator no significant changes occur within 10 days. The effect of hemolysis is slight. If 3 volumes of hemolyzed erythrocytes (calcium: 0.016 mmol/L) are mixed with 97 volumes of serum (calcium: 2.5 mmol/L) the calcium concentration in the serum falls to 2.43 mmol/L (–3%).

4.7 Ionized Calcium

Patient

The concentration of ionized calcium in the serum water may be slightly altered after food intake. The body position of the subject at blood collection does not have any influence on the serum concentration. A circadian rhythm with an amplitude of 0.07 mmol/L has been described. It is not synchronous with the circadian rhythm of total calcium.

Specimen

There are no differences between arterial and capillary blood, but slight differences between arterial and venous samples [123]. For determination of ionized calcium in the extracellular water phase of blood or plasma, anticoagulation with heparin in low concentrations (e.g. Vetren 2000, Promonta, Hamburg) is recommended in order to minimize binding of ionized calcium to the heparin. Otherwise false-low concentrations will be measured.

4.8 Ionized Phosphate

Patient

After food intake both a rise [55] and a fall in phosphate concentration are observed, possibly due to the varying composition of the food consumed or interference from the method. For this reason, fasting blood should be preferred for the determination of phosphate concentration in serum. The data on the influence of body position are contradictory. According to Röcker [91] one finds 2% lower values from supine persons compared to seated individuals, whereas Statland [104] reported 3.6% higher values. There is a circadian rhythm of phosphate concentration [15] and sampling should always therefore take place at the same time of day.

Specimen

No significantly different phosphate concentrations are found in serum samples from venous and capillary blood. The concentration in serum is about 0.08 mmol/L higher than in heparinized plasma due to cleavage of organic phosphate esters by erythrocyte enzymes. Plasma and serum should therefore be separated as soon as possible and may be stored for 24 h at 4 °C. During more prolonged storage, the phosphate concentration rises due to cleavage of organic phosphate esters. In urine samples, the phosphate concentration can rise on storage for the same reason, but may also fall due to precipitation of poorly soluble phosphate salts.

4.9 Electrolyte Determinations in Urine

24 h-Collected Urine

Electrolyte determinations in the urine are generally carried out on a 24 h-urine sample, because of the circadian variations in excretion. Before starting the collecting period, the bladder must be emptied and the resulting urine discarded.

During the collection period, the urine is to be collected in an adequately sized disposable container (e.g. 2 L plastic bottle) and stored in the dark. At the end of the collection period, the bladder must be emptied and the urine obtained added to the collected urine. Addition of preservatives is not necessary. With normal fluid intake, a urinary volume of 1.3–1.5 L/day will be measured [26].

Morning Urine

According to [60] the electrolyte concentration of the morning urine in healthy subjects approximately equates to the electrolyte concentration in 24 h-collected urine and can be used for estimating the amount of electrolytes excreted in 24 h. "Morning urine" is collected over a minimum period of 4 h.

4.10 Acid-Base Balance in Blood and Blood Gases

Blood samples for determination of blood gases and the acid-base balance can be influenced by a large number of factors. These factors must be taken into account if the blood gas analysis is to produce results which are a true and accurate reflection of the in vivo situation [21]. It is important to be aware that, while errors during the analysis itself can be detected by the laboratory's quality assurance measures, preanalytical errors can only be detected to a limited extent through plausibility checking (comparison with earlier values, consistency with other findings, examination of extreme values). Immediate measurement of the collected blood sample at the bedside (bedside testing, point of care testing) helps to avoid some preanalytical errors, e.g. those occurring during sample transport, on the other hand these tests are often performed by staff who have little experience in this field and using analyzers which for various reasons – e.g. technical design, available features, maintenance – are less reliable than the blood gas machines of the central laboratory.

Sample Containers

Glass containers (e.g. glass syringes or glass capillaries) are in principle gas-tight for at least 2 hours and therefore particularly suitable. Their disadvantage is the risk of breakage and injury and, in the case of glass syringes, the need for cleaning and sterilization to avoid infection. Plastic containers are more or less gas permeable which can lead to changes in pCO_2 and pO_2 in either direction, depending on the gas composition of the ambient air. However, if the samples are stored in ice water pCO_2

usually remains unchanged for 30 minutes, pO_2 below 200 mm Hg (27 kPa) for a maximum of 15 minutes.

Anticoagulants

Lithium-heparin is recommended as anticoagulant. The final concentration should not exceed 50 IU/mL blood (sodium heparin leads to (slightly) raised Na values at concentrations above 15 IU). If dry heparin is used dilution effects need not be taken into consideration. More anticoagulant is required for glass tubes than for plastic tubes on account of the rougher surface of the former. When using coated tubes attention must be paid that the tubes are completely filled, otherwise the concentration of anticoagulant will be higher than optimal and can lead to falsification of results:

100 IU/mL blood   pH   up to −0.0004
                  HCO_3^- up to −0.3 mmol/L
                  Base excess   up to −0.3 mmol/L

If heparin solution is used in place of solid heparin anticoagulation is achieved with particularly low concentrations:

Container Material	Heparin	Heparin Concentration
Glass	Dry substance	40–60 IU/mL
Glass	Heparin solution	8–12 IU/mL
Plastic	Dry substance	12–50 IU/mL
Plastic	Heparin solution	4–6 IU/mL

It is important to ensure that the heparin solution does not make up more than 5% of the total volume of the sample. Otherwise lower values for HCO_3^- and pCO_2 and higher values for pO_2 (as pO_2 of the heparin solution is 150 mm Hg (20 kPa)) will be measured commensurate with the dilution. In order to minimize the changes in the electrolyte concentrations, use of an electrolyte balanced heparin solution is recommended:

Sodium   120–150 mmol/L
Potassium   3.5–4.5 mmol/L
Chloride   100–130 mmol/L
Ionized calcium   1.2–1.4 mmol/L

Sample Collection

In order to obtain a representative blood sample the patient should have been in a respiratory steady state for 15–30 minutes. The sample must be collected anaerobically and anticoagulated immediately, taking care to avoid hemolysis. Contamination of the sample with infusion solutions must be avoided, preferably by puncture at a separate site, otherwise by discarding the first sample. After sample collection any air bubbles must be removed. The puncture needle is removed, the syringe tightly closed and the contents mixed by inverting and swirling.

Storage and Transport

If the analysis cannot be performed within 15 minutes the blood sample must be stored in ice water. Cooling slows down glycolysis which leads to a fall in pH, bicarbonate and base excess via formation of lactic acid.

Blood samples in glass syringes can be stored in ice water for at least 2 hours without changes in pH, pCO_2 or pO_2.

If pO_2 is also to be determined, blood samples in plastic syringes should not be stored in ice water but be kept at room temperature and analyzed within 15 minutes unless there is marked leukocytosis (more than 50 x 10^3 µL), thrombocytosis (more than 600 x 10^3 µL) or anemia (hemoglobin below 7.5 g/dL), in which case the analysis must be performed immediately. If only pH and pCO_2 are to be determined and not pO_2 the sample can be stored in ice water for 30 minutes.

After storage the sample must again be mixed by swirling before the measurement is performed.

4.11 Acid-Base Measurement in Urine

A 24 h-urine sample collected as described above is needed. On account of the particular danger of bacterial breakdown of urea to ammonia, addition of 10 mL thymol/isopropanol (1 + 9 v/v) to the vessel before collection is advised. After collection of the first urine portion the sample is overlaid with liquid paraffin. Later urine portions are introduced below the paraffin layer using a funnel.

5. Methods of Determination

The various electrolytes can be measured by means of electrochemical or spectroscopic methods (Tab. 52). The methods used for determining electrolytes in serum and plasma cannot be used directly for urine because, amongst other things,

- differing concentrations are present,
- the ratios of the electrolyte concentrations to each other differ and fluctuate markedly (e.g. sodium and potassium),
- the electrolyte may be present as a poorly soluble salt (e.g. calcium and phosphate),
- interfering factors may exist in other concentration ranges.

This means that, when using the same technique, other calibrators (sodium, potassium determinations) or pretreatment of the sample (calcium and phosphate determinations) may be necessary for urine.

Most of the following methods determine the molar concentrations of the analytes, e.g. the concentration of substance in 1 L serum and not the molal concentration in serum water. This is of importance for the medical interpretation of the values.

The monovalent ions of sodium, chloride and potassium are present almost exclusively in the aqueous phase of the serum or plasma. The methods, however, detect the concentration in the total serum, i.e. including the electrolyte-free compartment comprising the proteins, lipids and other macromolecules. The size of the electrolyte-free compartment is variable and depends on the space occupied by the macromolecules. The medical assessment is only possible when the "dilution" of the serum water by the macromolecules is "normal". Otherwise the values have to be corrected for interpretation [74]. If raised concentrations of glucose, urea, ethanol, etc. can be ruled out, then the osmolality determination can provide an indication of the altered size of the electrolyte-free space. Changes in the electrolyte-free space are medically relevant, especially for sodium ("pseudohyponatremia") and chloride because of their small relative reference interval, but are medically less important for potassium (large relative reference interval).

In serum about 40% of the calcium ions are protein-bound at normal protein concentrations. If a rise occurs in the protein concentration,

Table 52. Methods of determination of electrolytes in serum or plasma

Electrolyte		Spectroscopic Methods			Electrochemical Methods	
		Absorption spectrometry (AS)	Atomic absorption spectrometry (FAAS)	Flame atomic emission spectrometry (FAES)	Potentiometry with ion-selective electrodes (ISE)	Coulometry
Na^+	total	+		+	+	
Na^+	ionized[1]				+	
K^+	total	+		+	+	
K^+	ionized[1]				+	
Ca^{2+}	total				[+]	
Ca^{2+}	ionized[1]				+	
Mg^{2+}	total	+	+			
Mg^{2+}	ionized[1]				+	
Cl^-	total	+			+	+
Cl^-	ionized[1]				[+]	
PO_4^{3-}	total	+				

+ Possible, widespread use
[+] Possible, but not widespread use
[1] "Ionized" fraction of electrolytes is understood to mean a fraction that is neither bound to proteins nor complexed. The terms "ionized" and "free" are used synonymously in this connection.

the total calcium concentration increases, but the (medically relevant) concentration of free ("ionized") calcium remains constant. Correspondingly, when there is a fall in protein concentration a lowering of total calcium concentration is found (known as "pseudohypocalcemia", see section "Ionized calcium").

In fact in pathophysiology a raised or lowered electrolyte concentration generally signifies a pathological concentration of the free electrolytes in the aqueous phase of the respective system (in serum or plasma water or in the extracellular water phase of the blood).

Osmolality always refers to the aqueous phase. As a parameter for a colligative property, it has, however, only limited diagnostic specificity for the assessment of electrolytes.

5.1 Ion-Selective Electrodes

As early as the start of this century Cremer discovered that an electrical potential develops across thin-glass membranes of a given chemical composition if there are solutions with differing H^+-ion activities on the two sides of the membrane (Fig. 37). Intensive research with membrane electrodes in the last 25 years has brought electro-analytical methods without current flow to the fore – especially ion-selective potentiometry. In the physiological field, ion-selective electrodes (ISE) are the only practical method for determining ion activities (Tab. 53) [51].

Ion-selective electrodes and other probes fitted with specific membranes are being tested for continuous in vivo monitoring.

Table 53. Historical development

Year	
1906	Nernst equation
1938	Model of the membrane electrode (Cremer)
1967	Glass membrane electrodes for pH measurement and determination of cations (Eisenman)
1969	Use of specific ion carriers, e.g. valinomycin for K^+-selective electrodes (Simon)
1986	pH reference method (Maas et al.)

Construction of Ion-Selective Measurement Chain

To determine electrolytes by means of ion-selective electrodes (ISE) we use an ion-selective measurement chain as shown in Fig. 33. This consists of two electrodes, the ion-selective electrode and the reference electrode. The specific determination of a given ion species (e.g. chlorides) in the presence of other ions, and of a solution with a complex matrix, is achieved by the use of a specifically dedicated membrane. The reference electrode and its constant potential is necessary for deriving the specific potential difference. It is connected via an electrolyte bridge but with the measuring solution separated by a diaphragm.

On examination of blood, a suspension effect develops as the result of hemolysis and protein deposition at the separating layer between the measuring solution and the electrolyte bridge of the reference electrode.

Methods of Determination

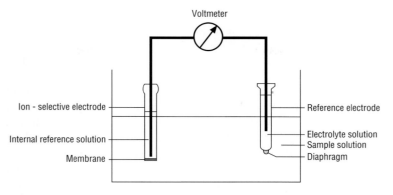

Fig. 37. Ion-selective measurement chain.

The measurement error caused by the suspension effect varies in intensity according to the construction of the reference electrode and the composition of electrolyte solution.

Nernst and Nikolsky Equations

An electrochemical cell consists of an ion-selective measuring electrode and a constant potential reference electrode. The measurable potential E of this cell depends on the activity of the free, unbound ions to be measured in the test solution. In the ideal situation, this dependence can be described in the form of an analytical function: the Nernst equation – if the residual potential (diffusion potential) and suspension effect can be ignored. The potential E that is measured is proportional to the logarithm of the relative molar activity of the ion to be determined.

$$E = E_0 + \frac{R \times T}{z \times F} \times \ln a_m$$

a_m	Relative molar activity of the ions to be determined = ion activity in the solution
E_0	Standard equilibrium potential ($a_m = 1$)
F	Faraday constant
T	Temperature in °K
z	Valency and sign of the ion to be measured

($R \times T$)/($z \times F$) is known as the Nernst factor and shows the temperature-dependent "slope" of the measuring chain.

The theoretical Nernst slope at 25 °C for monovalent ions is 59.16 mV/activity decade; for bivalent ions it is half ($z = 2$) and for trivalent ions ($z = 3$) a third of this value. The theoretical Nernst slope is seldom achieved by ion-selective electrodes. Downward deviations (about 90–98% of the theoretical value) are the general rule [22].

Reproducible deviations from the ideal Nernst behavior do not limit the analytical use of a measurement chain as the effective slope is taken into account in the calibration. Errors in measurement occur only if the slope alters in the interval between the calibration and measurement processes. Reasons for this may lie in the sample matrix, e.g. interfering ions that were not present in the calibration solutions.

The Nernst equation applies in cases where only one ion species contributes to the potential formation. We can then speak of an ion-specific electrode. In reality, however, ion-selective electrodes show varying degrees of sensitivity to other ions in addition to the ion of interest. Thus, a magnesium electrode will react with a certain degree of sensitivity to calcium and a potassium electrode to ammonium ions. This is taken into account in the Nikolsky equation.

$$E = E_0 + \frac{R \times T}{z \times F} \times \ln\left[a_i + \sum K_{ij}^{pot} \times a_j\right]$$

a_i Relative molal activity of the ion to be determined
a_j Relative molal activity of the interfering ion(s)

The parameter K_{ij}^{pot} is known as the potentiometric selectivity coefficient and shows to what degree the values measured with an ion-selective electrode designed for ion i are falsified by interfering ion j (or other interfering ions). An ideal and thus "ion-specific electrode" would have a value of $K_{ij} = 0$ for all other ions. Practical values for Kid lie between 10^{-13} ("alkali error" of a glass pH electrode) and 102 (cross sensitivity of a chloride-selective electrode to iodide ions).

The influence of the interfering ions arises from the fact that they too are able to pass the phase border and thus contribute to the formation of the potential.

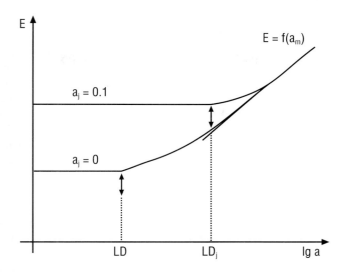

Fig. 38. Detection limit of ion-selective electrodes. aj: activity of interfering ion, am: activity of measured ion, E: measured potential, LD: limit of detection in the absence of interfering ions, LD_j: limit of detection in the presence of interfering ions.

Fig. 38 demonstrates this logarithmic relationship using the example of a sodium ion-selective electrode. From the figure it can be seen that the limit of detection (LD) is higher in the presence of interfering ions. If no interfering ions are present (e.g. 10^{-2} molar NaCl solution; $a_i = 0$) the limit of detection (LD) for sodium ions is about 2.5×10^{-6} mol/L. In the presence of interfering ions (here a) = 0.1 mol/L KCl) the limit of detection in the ion-selective potentiometry is higher (here, for example, about 1.5×10^{-5} mol/L = LD_j).

Ion-selective electrodes are not absolutely specific; they are more or less selective for a small group of ions. Errors caused by inadequate selectivity are difficult to estimate. Interference from other ions can best be determined by recovery experiments or checks with other analytical methods.

Relationships Between Activity and Concentration

In 1887 Arrhenius proposed the theory that conducting substances (salts, bases, acids) dissociate into ions in aqueous solution. In order to comply with universally valid laws (such as, for example, the law of mass action) it was necessary to describe ions and neutral molecules formally as being equivalent. In 1913, therefore, the activity coefficient was introduced as a correction factor by Lewis. The ion activity a_m ("active concentration") has since been defined as the product of the activity coefficient γ and ion concentration c.

$$a_m = \gamma \times c$$

As early as 1923, Debye and Hückel, in their theory of strong electrolytes, described the concentration dependence of the activity coefficients. Only in very highly diluted electrolyte solutions ($c < 10^{-4}$ mol/L) does the activity coefficient approach the value of 1 in which case ion activity and ion concentration are equivalent.

More recent investigations showed that the activity coefficient also depends on the charge of the ion and the concentration and charge of the other ions in the solution [84]. Measurement of an individual activity coefficient for a cation or an anion in a solution is therefore impossible. (Where this apparently takes place, e.g. in the definition of the pH, this is based on an international agreement.)

For work with ion-selective electrodes in serum or whole blood, the following differentiation between activity and concentration is especially important.

In infinitely diluted aqueous solutions of completely dissociated salts activity corresponds to concentration. With an increase in concentration, however, activity falls continuously as a result of electrostatic interactions. Thus, in serum only about 75% of the sodium ions are "active".

Complex bound ions are inactive and therefore not detectable by ISE.

Determination of activity refers – in the event of calibration with aqueous solutions – to the activity of the ions in the aqueous phase, even in investigations of samples of varying protein and lipid content. This means that the signal from the ISE is proportional to the active, ionized, free electrolyte in the aqueous phase.

Unless otherwise stated, concentration determination includes free and bound ions.

The concentration determination generally refers to the concentration in the overall system, e.g. serum. Where an increase occurs in the electrolyte-free space through an increase in macromolecules such as lipids, there is a reduction in ion concentration (see "pseudohyponatremia"). This means that, in concentration determination, the measuring signal is proportional to the stoichiometric total amount of electrolytes in the sample, including its electrolyte-free compartment.

For the reasons given the conversion of activities into concentrations is only possible on the basis of a number of (untestable) assumptions and, strictly speaking, is not permissible.

In investigations of samples by means of ion-selective electrodes one must differentiate between measurements on undiluted samples and measurements after dilution of the sample. The analysis of undiluted serum is a measure of the ion activity of the native material.

With high predilution of the serum, the ion strength of the sample is generally adjusted to the ion strength of the calibration solution, and the electrolyte-free compartment of macromolecules is reduced to less than 1% of the total volume. Using this measurement technique the measurement signals are converted, by comparison with calibration solutions, into concentrations. The high dilution does not allow any deductions to be made about the ion activity of the native sample. The concentration values obtained are identical to the results obtained with, e.g., flame atomic emission spectrometry if the bound fraction is negligible. This means that the accuracy of the analyses with ion-selective electrodes after dilution of the sample can be checked by means of reference method values [67].

The quantity of "direct" ISE (i.e. measurement in undiluted samples) is the ion activity of the free ions of a given species in the aqueous fraction of, for example, the plasma. Cells, proteins and lipids do not influence the measurement as the electrolytes are dissolved only in the plasma water. The "indirect" measurement methods, e.g., of the unbound alkali ions with prior dilution (flame photometry and ISE), by contrast, determine the concentration, i.e. the total amount of material per volume. The total volume of a plasma sample normally consists of 93% water and 7% proteins and lipids. In certain pathological conditions (dehydration, hyperproteinemia, hyperlipidemia) the aqueous fraction can, however, be considerably reduced (down to 85%), so that the value determined in such a sample by flame photometry is lower than normal in comparison with that of direct ISE.

This effect means that "direct" potentiometry, on the one hand, and "indirect" potentiometry or flame photometry, on the other, provide numerically different data in addition to the different quantities. To avoid a possibly fatal confusion in the handling of the two quantities, and to prevent the expansion of the flood of reference interval tables, the International Federation of Clinical Chemistry (IFCC) is about to propose a pragmatic convention for dealing with ISE values for sodium and potassium [17].

- The results of direct ISE measurements in whole blood or undiluted plasma should be given in concentration units (mmol/L), although activities are measured.
- The results of measurements of sodium and potassium in standardized normal plasma (serum) samples obtained with ISE and flame photometry should agree exactly.
- Standardized normal plasma (serum) means a plasma water fraction of 0.93 kg/L, a bicarbonate concentration of 24 mmol/L, a pH of 7.40 and concentrations of albumin, total protein, cholesterol and triglycerides within the reference range of healthy subjects.

Calibration of ISE with primary standard solutions will give the unbound sodium concentration in serum water. It is shown by Fig. 39 that with calibration according to the new convention in hyperproteinemia the differences between the values measured with flame photometry and "direct" ISE are reduced, but in hypoproteinemia the differences are increased. Only the protein concentration at which both methods yield the same value is shifted. Nevertheless, the "direct" ISE still determines a different quantity as compared to other methods, and the values obtained show differences at pathological serum water concentrations [64, 65, 73, 76].

From the medical point of view it is desirable that determinations in serum, plasma or blood should be undertaken without dilution of the native material so as to determine the concentration in serum water, plasma water or extracellular water phase of blood. These can be interpreted directly without knowing the protein and lipid concentrations or

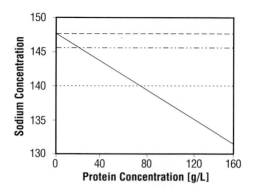

Fig. 39. Relationship between protein concentration and sodium concentration in serum and serum water respectively. —— Total sodium concentration in serum [mmol/L]; total molar concentration. ········ "Ionized" sodium (IFCC) [mmol/L]. —··—·· Free sodium concentration in serum water [mmol/kg]; free molal concentration. ----- Total sodium concentration in serum water [mmol/kg], total molal concentration. IFCC: International Federation of Clinical Chemistry.

the degree of complex formation. The possible use in this case of (anticoagulated) blood permits especially rapid determinations, in some analytical devices jointly with blood gas analyses.

Determinations in urine, generally carried out for calculating electrolyte balances, should be undertaken by means of flame atomic emission spectrometry or, in a second line, by means of ISE after dilution of the samples to determine the total concentration.

Indirect and Direct Potentiometry – Determination of Sodium, Chloride, Potassium, Magnesium, Calcium with ISE

"Ionized" and Total Sodium

The sodium determination is carried out with an ion-selective electrode which generally contains a sodium-sensitive glass electrode. However, for a number of years, membranes with ion carriers (e.g. ETH 227, ETH 157, methylmonensin) have also been used.

Sodium Determinations Without Dilution of the Sample (Ionized Sodium)

On calibration with aqueous standard solutions, the measurement signal is proportional to the relative molal activity of sodium ions a_{Na} in the aqueous phase (e.g. serum water). pNa can be defined analogously to pH:

$$pNa = -\log a_{Na}$$

Assuming that the activity coefficient on which the calibration is based and that of the sample coincide, then the free molar concentration m_{Na} can be calculated (reference interval: 142.5–153.1 mmol/kg). Generally, the activity coefficient of sodium in serum fluctuates within narrow limits by about 0.747. The bound fraction of the pertinent ions is not included in measurements with ion-selective electrodes. For alkali ions this is a small amount, e.g. for sodium in the serum normally approximately 2 mmol/L (1.5%, see Fig. 5).

The report of the values of ion-selective electrodes as total molal concentrations is erroneous to a greater or lesser degree, depending on the extent to which the actual activity coefficient and the actual binding deviate from the assumed values [17, 62, 63].

$$m_{tNa} - m_{Na} + m_{NaProt} + m_{NaX}$$

m_{Na}	Free molal sodium concentration
m_{tNa},	Total molal sodium concentration
m_{NaProt}	Molal sodium concentration bound to proteins
m_{NaX}	Molal sodium concentration bound to anions (e.g. HCO_3^-)

Reference interval for the total molal sodium concentration: 144.7–155.4 mmol/kg.

The differences between free and total molal concentration in the sera of patients are generally low, but large deviations may occur in control sera.

The molar concentration of sodium in serum C_{tNa} is calculated as follows:

$$C_{tNa} - (m_{Na} + m_{NaProt} + m_{NaX}) \times \rho H_2O$$

ρH_2O Mass concentration of water in serum (kg/L).

Whilst the activity coefficient and the binding of sodium generally deviate only slightly from the assumed values, the water concentration can fluctuate to a considerable degree, e.g. in hyperproteinemia, hyperlipidemia, administration of plasma expanders, but also in hypoproteinemia. In a group of hospitalized patients pathological protein concentrations were found in about 40% of cases. Conversion of free molar sodium concentration (mNa) to molar concentration (Ct_{NA}) without knowledge of the water concentration (and the binding fraction to proteins, etc.) can therefore lead to considerable errors. It is therefore preferable to report the values that are obtained by determination with ion-selective electrodes, without dilution of the sample, as pNa or free molal concentration in serum water. Stating the result as molar concentrations in the serum, considerable deviations from the correct results may occur because of unfulfilled assumptions [74] (see Tab. 62). A great deal of uncertainty can develop, especially if, alternately (analytically correct) concentrations obtained with flame atomic emission spectrometry are reported to the attending physician. It is therefore advisable to identify the "concentrations" calculated by means of ISE without dilution of the sample.

The differences between the correct concentration determinations of sodium in the serum by means of, for example, flame atomic emission spectrophotometry and the incorrect concentration determinations by means of ion-selective electrodes (ISE) without sample dilution are less if the ISEs are calibrated with protein-containing samples of known sodium concentration. For this purpose, the National Institute of Standards and Technology (NIST, previously NBS) has produced corresponding calibration material which is currently being tested (standard reference material (SRM 956) with three different sodium and potassium concentrations). It is to be expected that the use of this material will give correct concentration values in sera with "normal" water concentrations, normal activity coefficients and normal fraction of complex-bound sodium. As, however, these preconditions are frequently not present and in individual cases are not ascertainable, one will be obliged nevertheless to report separately the findings obtained by means of ion-selective electrodes without dilution of the sample (e.g. "ionized" sodium). Analytically these results are extremely unsatisfactory as they cannot be assigned throughout to clearly defined measurement parameters. Some results will correspond in practice to the correct concentration in the serum, but many others will show distinct deviations. The medical interpretation of these findings has the advantage that they can be compared without restriction to the corresponding reference interval. By contrast, the reference intervals for the analytically correct concentration determination apply only to samples with "normal" water content. In hyperproteinemia, for example, the reference interval is correspondingly lower and, in hypoproteinemia, higher.

Sodium Determinations After Dilution of the Sample (Total Sodium)

If the samples (serum or plasma) are diluted before measurement with an ion-selective electrode, then, as with flame atomic emission spectrometry, one obtains the correct molar concentration of the sample C_{tNa} (Fig. 40).

"Ionized" and Total Chloride

Various selective electrodes are used for chloride determination: silver chloride – solid state electrode, membrane electrodes with ion exchangers (e.g. methyltridodecylammonium chloride) or with neutral carriers.

At present total chloride concentration is generally determined by diluting the material to be examined (serum or plasma) before measurement. In practice, one can dispense with the determination of active chloride molality (by means of ion-selective electrode without sample dilution). With some electrodes a marked bromide interference is observed, which may even lead to a negative anion gap ("pseudohyperchloremia") [34].

"Ionized" and Total Potassium

A valinomycin-containing membrane within an ion-selective electrode is generally used for potassium determination.

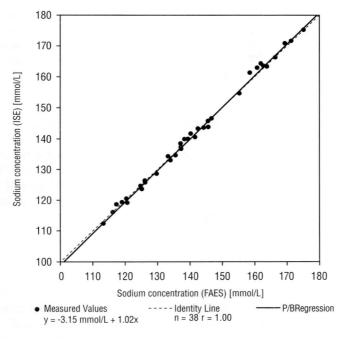

Fig. 40. Comparison of sodium determination in serum by means of flame atomic emission spectrometry (FAES) and ion-selective electrodes (ISE) after dilution of the sample. x-axis: sodium concentration in serum (mmol/L) by means of flame atomic emission spectrometry (FAES), y-axis: sodium concentration in serum (mmol/L) by means of ion-selective electrodes (ISE) after dilution of the sample (Roche/Hitachi 911), P/B: calculation according to [7].

Potassium Determinations Without Dilution of the Sample

These activity determinations are generally carried out simultaneously with activity determinations of sodium. Reporting of the results should therefore take place in the same manner as described under "sodium". The influence of the water concentration of the sample on the calculation of the total potassium concentration in the serum is just as large as with sodium.

The medical relevance is, however, less significant because of the greater relative reference interval. With an increase in the electrolyte-free compartment of 5%, one obtains, for example, 4.3 mmol/L instead of 4.5 mmol/L (in comparison with sodium where one obtains 133 mmol/L instead of 140 mmol/L).

Potassium Determinations After Dilution of the Sample

The (correct) molar potassium concentration is obtained if the analytical material (plasma or serum) is diluted before measurement. The differences between measurements without, and after dilution of the samples are less important medically for potassium than for sodium, even though analytically of the same magnitude.

"Ionized" Magnesium

The determination of magnesium by means of ion-selective electrodes is in its early stages. Membrane electrodes containing neutral carriers (e.g. ETH 5220 or ETH 7025) are used. They are not strictly specific for magnesium, but are also sensitive to calcium ions. For this reason the calcium ion activity is simultaneously determined separately and its influence on the magnesium determination compensated mathematically [6, 70]. For determination one has to take account of the fact that 30% of magnesium ions are bound to protein [70]. This figure is increased in alkalosis and reduced in acidosis (see "ionized" calcium). In case of liver transplantation ionized magnesium concentration in the serum may be extremely reduced by complexing with citrate [69, 71].

"Ionized" Calcium

The ion-selective electrodes for determining ionized calcium consist of the Ca^{2+}-selective membrane electrode and the reference electrode. The membrane electrode generally contains an ion carrier (e.g. ETH 1001). The undiluted sample is always used for the determination. $CaCl_2$ solutions with an ionic strength of 160 mmol/kg by addition of NaCl are used for calibration purposes. This means that the activity of the calcium ions in the aqueous phase of the sample is determined by comparison with the signal obtained for the calibration solution, and the concentration of "ionized" calcium in the aqueous phase of the sample (e.g. plasma water) is calculated on the assumption that identical ionic strengths exist in the sample and calibration solutions, and that therefore the same activity coefficients apply to the ionized calcium in the sample and calibrator. According to Siggaard-Andersen [96] the concentration of ionized calcium measured by this procedure at a sodium concentration of 170 mmol/L is too low for calcium by 0.11 mmol/L and, at 110 mmol/L, is 0.10 mmol/L too high in comparison with the sodium concentration of 140 mmol/L, as the activity of the calcium ion decreases or increases under these conditions. Fluctuations in the sodium concentration within the reference interval are, by contrast, negligible for the concentration determination of ionized calcium by means of ion-selective electrodes.

The binding of calcium to protein, for example, is pH-dependent. Only anaerobically collected samples therefore can reflect the activity or concentration of ionized calcium in vivo. However, no reference intervals are available for values deviating from pH 7.4. If a normal acid-base balance can be assumed, then the sample may be collected aerobically. In case of storage (with loss of CO_2) it is adjusted with CO_2 to pH 7.4 ± 0.2 before measurement. The values obtained are adjusted, with the aid of a suitable algorithm, to pH 7.4 and these can then be compared with the reference interval.

The measurement of ionized calcium has generally gained preference over its calculation with the aid of nomograms, proceeding from total calcium concentration and taking into account, for example, albumin and protein concentrations and pH. One of the reasons is that, in mass transfusion, the binding behavior underlying the calculations no longer applies as a result of the addition of citrate, so that, with raised total calcium concentrations, the concentration of ionized calcium may be reduced [43]. Furthermore, the determination by means of ion-selective electrodes has since become so reliable and practicable that it is now preferable to a calculation procedure which requires the determination of three different quantities.

The determination of total calcium by means of ISE, a procedure in which the bound calcium must first be released from its binding, has so far not experienced widespread use.

5.2 Absorption Spectrometry – Photometric Determination

The preferred measuring techniques for determining most electrolytes are at present (still) flame atomic emission spectrometry (FAES), flame atomic absorption spectrometry (FAAS) and potentiometry by means of ion-selective electrodes (ISE), as well as coulometry for chloride determination (see Tab. 52). As the analyzers used in clinical chemistry invariably contain an absorption spectrometer as a central module, determining electrolytes by absorption spectrometry is the obvious option. This makes it possible to do without additional instruments such as FAES, FAAS or coulometry, which simplifies instrument handLing and generally enables a higher sample throughout per unit of time [120]. Developments in recent years have produced major advances in this area (Tab. 54).

Methods in which electrolytes are determined "enzymatically" are especially worthy of mention. In these procedures the electrolytes can act, for example, as effectors in an enzyme-catalyzed reaction in which the corresponding substrate is converted more or less rapidly depending on the concentration of the effector. The methods can be readily adapted to automated analytical devices, require only small sample volumes (2 µL to 20 µL) and, moreover, possess good precision, accuracy and specificity, both for serum and plasma, not to mention urine samples.

The dependence of enzyme activities on ion concentrations opens up the fundamental possibility of converting electrolyte determinations to simple enzyme activity determinations.

Numerous enzymes activated by monovalent ions have been described. The following criteria must be taken into account:

- Highest specificity for the ion to be determined.
- Linear dependence of the enzyme activity on the concentration of the ion to be determined (in a concentration range relevant for the measurement).
- Adequate stability even in the absence of activating ions.
- Simple photometric activity determination.
- Availability in adequate quantity and purity.
- No interference by any drugs or metabolites that may be present in the serum.

Table 54. Historical development

Year		References
1956	Absorption spectrometric Mg^{2+} determination with xylidyl blue	[80]
1964	Absorption spectrometric Ca^{2+} determination with o-cresolphthalein	[56]
1967	Enzymatic phosphate determination	[79, 94]
1985	Enzymatic Mg^{2+} determination	[39, 106]
1988	Enzymatic Na^+ determination	[11, 12, 89, 110]
1988	Enzymatic Cl^- determination	[87]
1989	Enzymatic K^+ determination	[13]
1994	Enzymatic Ca^{2+} determination	[54]

As these "enzymatic" procedures have not become established in routine practice they are not given further attention here (Tab. 54) [52].

Magnesium

A widely used absorption spectrometric method for the determination of magnesium in serum or urine is based on the reaction of magnesium with xylidyl blue in alkaline solution containing EGTA to mask the calcium in the sample. The change in absorbance arising from complexing of magnesium by the reagent can be measured at wavelengths of 600/505 nm on an analyzer, e.g. Roche/Hitachi 911 (Roche Diagnostics GmbH) [32, 80].

The change in absorbance is based on the properties of the diazonium dyes (xylidyl blue) in forming stable complexes with Mg^{2+} in the alkaline pH range. The complex formation is accompanied by a shift in the absorbance bands to shorter wavelengths. The absorbance peak of xylidyl blue is at 620 nm. Complexing with magnesium produces a shift of this peak to 520 nm. The method is preferred to the determination of magnesium by calmagite (Fig. 41).

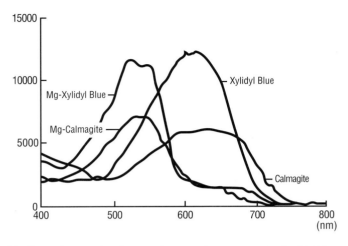

Fig. 41. Absorption spectra of magnesium complex compounds of xylidyl blue and calmagite.

Calcium

The photometric methods for determination of total calcium in serum, urine and other body fluids are based on the formation of colored calcium complexes with chloroanilinic acid, calcein, o-cresolphthalein and others. Direct and indirect fluorimetric procedures for calcium determination are used by only a very few laboratories [37].

The currently preferred, and almost exclusively used, absorption spectrometric method for determination of calcium in serum is the reaction with o-cresolphthalein using automated analyzers. In alkaline solution Ca^{2+} combines with o-cresolphthalein to form a violet complex, which is measured at 546 nm. Protein-bound calcium is released with hydrochloric acid and interference by magnesium is prevented by means of 8-hydroxyquinoline complexing.

Phosphate

Various methods for absorption spectrometric determination of inorganic phosphate are used in clinical chemistry laboratories. The preferred method is the reaction of phosphate with ammonium molybdate in which a mixture of polyacids of complex structure is produced. Frequently, the polyacids are reduced by a reducing agent [e.g. tin(II) chloride] to molybdenum blue. In the determination of phosphate by use of ammonium molybdate falsely elevated values (pseudohyperphosphatemia) are found if removal of protein is omitted in hyperproteinemia (e.g. as a result of multiple myeloma). However, reagents have become commercially available (Roche Diagnostics GmbH) which eliminate this interference. In addition the reaction of phosphate with vanadate is also used for the determination of phosphate concentration. However, this method tends to overestimate inorganic phosphate due to hydrolysis of organic phosphate esters [36].

5.3 Flame Atomic Emission Spectrometry ("Flame Photometry"; Sodium, Potassium, Calcium)

Flame atomic emission spectrometry (FAES) is a reliable method for the determination of sodium, potassium and calcium in biological fluids, as well as for the determination of lithium in serum in the course of monitoring of lithium salt therapy. The procedure is less reliable for calcium determination than atomic absorption spectrometry. The reference methods for the determination of sodium [114] and potassium [115] in serum are based on FAES. During routine use of FAES the precision, and especially the accuracy, of the reference method are not achievable. The introduction of flame atomic emission spectrometer (flame photometer) was undoubtedly one of the most important developments in clinical chemistry in the last 60 years (Tab. 55). The

Table 55. Historical development

Year		Reference
1860	Introduction of emission spectroscopy in inorganic analysis (Kirchhoff and Bunsen)	[57]
1873	<Spectronatromètre> for quantitative determination of sodium (Champion et al.)	[24]
1934	Construction of a continuous nebulizer for flame atomic emission spectrometer (Lundegårdh)	[75]
1978/1979	Reference methods for the determination of sodium and potassium in serum (Velapoldi et al.)	[114, 115]

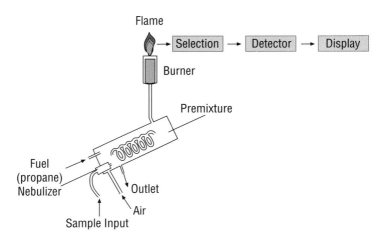

Fig. 42. Essentials of a flame atomic emission spectrometer ("flame photometer").

rapid, reliable determination of sodium and potassium that was thus made possible was decisive in improving the diagnosis and therapy of electrolyte disorders.

In FAES the diluted sample is sprayed by means of a nebulizer as a fine aerosol directly or indirectly into a flame (Fig. 42). A series of complex processes then occurs. The salt molecules dissociate into ions, some of which take up electrons and become free atoms, which in turn can convert to an excited state through uptake of thermal energy. The flame therefore contains a mixture of molecules, atoms, excited atoms and ions. Some of the essential processes are shown in Fig. 43.

In flame atomic emission spectrometry (FAES) only reaction step (4) in Fig. 43 is measured. The metal atoms take up a discrete energy portion (3) and emit the absorbed energy as light of a defined wavelength (Na: 590 nm, K: 767 nm). The intensity of the emitted radiation is proportional to the concentration of the ions in the sample. Relatively narrow bands ("lines") are typical for the emission of light by atoms. The emission bands for sodium, potassium and calcium are shown in Fig. 44.

The excitation energy for alkaline earth metals, and especially for alkali metals, is low (as these only have one electron on the outer shell). Only about 1–5% of the atoms are in excited state at any time. The ions

(1)	AX + T.E.	→ A$^+$ + X$^-$
(2)	A$^+$ + e$^-$	→ A$^\circ$
(3)	A$^\circ$ + T.E.	→ A*
(4)	A*	→ A$^\circ$ + E

Fig. 43. Excitation and decomposition processes in the flame. A$^+$: Metallic ion, A$^\circ$: metallic atom, A*: excited metallic atom, e: electron, E: energy of the emitted light quanta, T.E.: thermal energy, X$^-$: non-metallic ion.

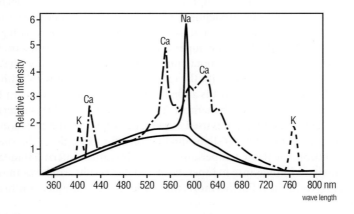

Fig. 44. Emission bands of sodium, potassium and calcium atoms.

that arise with increasing flame temperature can also be stimulated to emit light. This leads to an ion spectrum which differs from the atomic band spectrum. The experimental conditions must be adapted to ensure that the light emission of the ions is as low as possible and constant. This is achieved by using a flame temperature that is not too high.

With samples with relatively constant composition – for example serum – reliable results are obtained. For analyses of materials with markedly fluctuating sodium and potassium concentrations, e.g. urine, larger deviations in accuracy are to be expected.

By setting a suitable wavelength using a suitable selecting device (monochromator or a filter with a low half band width) spectral interference can be avoided for sodium and potassium determinations. For determinations of sodium and potassium in serum, cationic and anionic interference are of no importance. The anionic interference which occurs in calcium determination because of poorly evaporable phosphate salts can be managed by the addition of excess phosphate or the use of a higher flame temperature.

Use in Clinical Chemistry: "Flame Photometer"

In clinical chemistry FAES employing an internal standard is preferred. Either lithium (lithium guideline) or cesium are recommended as internal standards. On thermic excitation both elements emit photons at given wavelengths which do not superimpose over the irradiations of, for example, excited sodium or potassium atoms and they are only present in traces in patients' sera (with the exception of treatment with lithium salts). The internal standard is added at the same concentration to the calibration solution and the sample. The light emission of this internal standard gives a signal that is used as reference signal for the emission of the respective analytes. As slight fluctuations, for example in the speed of aspiration, the flame temperature or stability, affect internal standard and analytes in the same manner, fairly reliable results can be obtained with this technique. Furthermore, the addition, for example, of lithium ions inhibits the interfering ionization of potassium atoms.

Calibration solution and control material should be as similar as possible to the analytical sample as regards the concentration of the various ions. The viscosities should be equalized by dilution of the sample in order to obtain as identical flow and aerosol properties as possible. This means that it is not the analytical technique alone (FAES) that ensures reliable results, but also the manner in which it is executed.

Flame atomic emission spectrometry can also be used for the determination of sodium and potassium in urine. Here, however, different calibrators should be used, which may often differ considerably in composition from the urine samples in question, as the concentrations of sodium and potassium in the urine can fluctuate markedly – in contrast to serum. The accuracy of urine analyses may therefore be poorer than that of serum analyses.

5.4 Atomic Absorption Spectrometry (Magnesium, Calcium)

Atomic absorption spectrometry (AAS) is a universally utilizable procedure with which over 60 different elements can be determined [55, 125]. It is used in clinical chemistry for quantifying calcium, magnesium and lithium, as well as copper, zinc, lead and other trace elements. The reference methods for determination of calcium [16], magnesium [68] and lithium [117] are based on AAS.

Table 56. Historical development

Year		Reference
1860	Experimental elucidation of the relationship between emission and absorption (Kirchhoff)	[57]
1955	Procedure for quantitative element determination (Walsh)	[121]
1970	Atomization of organic samples in graphite tube furnaces (Massmann)	[82]

The fundamental relationships in AAS were elucidated as early as 1860 by Kirchhoff and Bunsen through their investigations on line inversion in alkali and alkali earth spectra, and formulated in the law "Each substance can absorb the radiation of the wavelength which itself emits" (Tab. 56). Thus, absorption lines of individual elements, e.g. sodium, can appear in the continuous emission spectrum of the sunlight as interruptions (black lines). The emission line of sodium lies in the same position of the spectrum as the black D-line in sunlight (Fraunhofer's line).

The construction of a flame atomic absorption spectrometer is shown diagrammatically in Fig. 45.

- A hollow cathode lamp emits photons of the line spectrum of the element to be determined.
- The sample to be examined is nebulized and then evaporated in the burner and its inorganic molecules are dissociated into atoms. In the hot atomic cloud the element-specific photons (emitted by the hollow cathode lamp) are absorbed by the non-excited atoms of the element to be determined.
- After isolation of the desired wavelength by means of a monochromator grating the attenuation of the radiation of the resonance wavelength is measured in a detector [125].

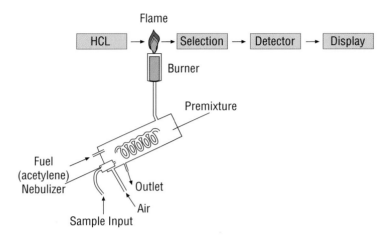

Fig. 45. Essentials of a flame atomic absorption spectrometer (FAAS).

AAS is a technique in which only one element is determined at a time. For analysis of another element the hollow cathode lamp must be changed, and the wavelength and several working parameters optimized.

The Bouguer-Lambert-Beer law applies, i.e. there is a linear relationship between absorption and concentration of atoms in the aerosol. Experimentally the concentration is determined with the aid of calibration solutions containing known concentrations of the element. Concomitant substances that are present in the sample in addition to the element to be determined may cause interference, which can lead to incorrect measurements. The interference may be attributable to spectral or non-spectral interference.

- Excessively high values are falsely measured with background absorption, i.e. with absorption or radiation by molecules of gaseous substances or by radiation scatter. Spectral interference caused by similar or identical absorption lines of other elements are very rare.

Non-spectral interferences include the differing behaviors of the sample and calibrator during vaporization and the gas phase.

- With low dilution of the sample, the viscosity of the aqueous calibration solutions must be adjusted to the biological fluid, e.g. by addition of gelatin.
- The alkaline earth metal ions Mg^{2+} or Ca^{2+} form stable salts with phosphate, with the result that the formation of the free atoms in the ground state used for FAAS is restricted to varying degrees. Lanthanum(III) chloride is used as a "phosphate trap" to avoid this "anionic interference".
- In the flame there is equilibrium between the ionized and the non-ionized metal atoms. Ionization increases with increasing flame temperature. By the addition of a sufficiently large amount of another readily ionizable metal salt to the sample and calibration solution, this equilibrium is adjusted in favor of the non-ionized form of the analyte and stabilized. The method thus permits accurate and sensitive determinations.

The technique gives reliable results only if the measurements are carried out carefully with attention to possible interfering factors, i.e. the method is one for qualified personnel.

5.5 Coulometry (Chloride)

Coulometry is used in clinical chemistry exclusively for the determination of chloride in serum or urine, both as a routine method [55] and as a reference method in a more sophisticated manner [116].

A given aliquot of the material to be examined is pipetted into a special container. On starting the device a silver wire (anode) in the generator circuit is subjected to a constant current and, in accordance with Coulomb's law, silver ions are released. These precipitate chloride ions in the solution as AgCl until all halogen ions are bound as silver halogenide. In the indicator circuit the current suddenly increases at the point at which free silver ions appear, marking the end of current flow. The duration of the current flow in the generator circuit is a measure of the quantity of chloride ions in the material under investigation, insofar as other halogen ions are not present. False chloride determinations may rarely occur, for example during chronic administration of bromide-containing drugs – because of the long bromide half-life. 1 mmol bromide is incorrectly measured as 1 mmol chloride.

5.6 Osmometry (Osmolality)

Osmometry is used to determine the osmolality of serum and urine, and is necessary for calculating free water clearance and the osmotic gap.

When 1 mol of particles is contained in 1 kg of water, then the freezing point falls by 1.858 °C, i.e. the freezing point depression is a measure of the quantity of particles contained in the sample (cryoscopy). The sample is supercooled to a given temperature. A vibration impulse initiates crystallization and heat is liberated. The temperature rises to the freezing point equivalent to the number of particles. With adequate calibration the result can be read directly in mmol/kg water.

One can also utilize the fact that, in the presence of 1 mol particles/kg water, the boiling point of the water increases by 0.52 °C. In this procedure, however, only osmotically active nonvolatile substances are determined.

5.7 Carrier-Bound Reagents ("Dry-Phase Technology")

In "dry-phase technology" carrier-bound reagents (preportioned reagents) are used which follow methodological principles that are also common in "traditional" clinical chemistry. Potentiometric measurements and color reactions are used for electrolyte determination. Serum or plasma may be used as specimen and, for some strips, blood as well (Reflotron®, Roche Diagnostics GmbH) [104].

Sodium

The sodium determination can be carried out potentiometrically using Vitros (Ortho). The values are equal, apart from paraproteinemic sera, to the sodium concentrations in serum ("flame photometry") and differ from the values obtained by ion-selective electrodes without sample dilution [61, 63]. Determinations in urine are also possible by using a special dilution solution.

Chloride

Chloride can be determined potentiometrically with carrier-bound reagents (Vitros, Ortho). With normal protein concentrations the values agree, with chloride concentrations determined, for example, by coulometry; in hyperproteinemia higher values are found, as expected [61, 63]. A special procedure is recommended for chloride determinations in urine.

Potassium

Various test-strips are available for potassium determination.

1. Reflotron® (Roche Diagnostics GmbH): Potassium in the applied blood sample is bound by valinomycin. The proton released in exchange leads to a color change in an indicator [118].
2. Vitros (Ortho): The values obtained by means of potentiometry correspond to potassium concentrations in the serum, apart from paraproteinemia [61, 63]. The slides can also be used for potassium determination in urine, but the samples must be mixed beforehand with a special dilution solution.

Magnesium

Protein-bound magnesium is released and reacts with the indicator dye. Calcium interference is prevented by the addition of a chelating agent (Vitros, Ortho). The method is less accurate in the analysis of control samples than in native samples [63]. Urine samples can also be analyzed by prior dilution and adjustment to pH 3–4.

Calcium

First of all, calcium is released from its protein binding after application to the "slide". The electrolyte migrates through the various layers and binds with the indicator dye arsenazo III (Vitros, Ortho). Accuracy of the test when used with native sera is good, but in control samples there more distinct deviations occur – probably matrix-induced [63]. Urine samples can also be analyzed using the slide.

Phosphate

The reaction with ammonium molybdate with subsequent reduction is used for phosphate determination (Vitros, Ortho). In hyperproteinemia false-high values are measured [63]. Urine can also be examined after about 10-fold dilution.

6. Methods of Determination of Acid-Bases and Blood Gases

6.1 Acid-Base Balance in Blood

pH

The pH is defined as the negative logarithm of the relative molar hydrogen ion activity. It is thus a dimensionless quantity.

The pH is derived directly from the Nernst equation. It is possible to convert the result (approximately) into the hydrogen ion concentration but this is rarely done. The pH decreases with increasing temperature and is determined with so-called blood gas analyzers at 37 °C. Conversion to other temperatures is possible. The reference method with the least uncertainty is based on the so-called Harned Cell which can, however, only be used for dilute aqueous solutions containing only inorganic components [8, 30]. The IFCC reference method [77] on the other hand is suitable for blood and other biological fluids. This method uses a pH glass electrode and a calomel reference electrode containing saturated KCl solution as electrolyte bridge. Calibration is performed with phosphate buffers certified by the National Institute of Standards and Technology (NIST), Gaithersburg, using the Harned Cell. The method used by routine blood gas machines is similar to the IFCC reference method. A pH selective glass electrode filled with a solution of known pH (internal reference solution) is used. On contact with aqueous solution the glass surfaces gradually form gel layers which take up protons if the test solution has an acid pH or give off protons if the pH is alkaline. The test solution remains negatively or positively charged, resulting in a potential difference which – the pH of the internal reference solution remaining constant – reflects the pH of the test solution. An Ag/AgCl electrode is usually used as reference electrode. It is calibrated with phosphate buffers the pH of which has been certified using NIST buffers. If ionized calcium, for example, is to be determined at the same time, a different buffer solution must be used (e.g. HEPES buffer system) in order to avoid an interaction between PO_4^{3-} and Ca^{2+}.

If the pH increases by 1 (e.g. pH 7 to pH 8) the potential theoretically decreases by 61.54 mV (37 °C). In practice the value should not be more than 4% below the theoretical value.

The partial pressure of a gas in a solution is defined as its partial pressure in an ideal gas which is in steady state with the solution [99]. The partial pressure is still usually given in mm Hg or Torr although the recommended SI unit is kPa (1 mm Hg corresponds to 0.133 kPa).

No reference method for determination of pCO_2 (CO_2 partial pressure) has been specified. However, as ideal reference material can be produced by tonometry, each laboratory can produce its own reference material for checking the accuracy of its measurements [19].

The pCO_2 is determined using a glass measuring electrode which is filled with a bicarbonate solution and separated from the test solution by a membrane which is permeable only for CO_2. Depending on the pCO_2 of the test solution more or less CO_2 molecules enter the bicarbonate solution, which results in a change in the pH. This pH is a measure of the pCO_2 of the test solution:

$$K = \frac{[H^+] \times [HCO_3^-]}{[H_2CO_3]}$$

$$pK = pH - \log[HCO_3^-] + \log[H_2CO_3]$$

$$pH = pK + \log \frac{[HCO_3^-]}{[H_2CO_3]}$$

If $[H_2CO_3]$ is replaced by pCO_2 and α_{CO_2} this gives the Henderson-Hasselbalch equation:

$$pH = pK + \log \frac{[HCO_3^-]}{pCO_2 \times \alpha_{CO_2}}$$

α_{CO_2}: molar solubility coefficient for CO_2 (0.0307 mmol/(L mm Hg) or 0.230 mmol/(L kPa)).

Since pK, α_{CO_2} and $[HCO_3^-]$ (which is present in great excess) can be regarded as constant, it follows that the pH reflects the pCO_2:

$$pH = -\log pCO_2 + \log[HCO_3^-] - \log \alpha_{CO_2} + pK^+$$
$$pH = -\log pCO_2 + C$$
$$C = \log[HCO_3^-] - \log \alpha_{CO_2} + pK$$

pK^+, the negative exponent of the apparent constant of the steady state $CO_2 + H_2O \leftrightarrow H^+ + HCO_3^-$ is used instead of pK.

pK^+ is not a thermodynamic constant but is dependent, inter alia, on pH and ionic strength. Nevertheless a constant value of 6.105 (in plasma) – or 6.095 in blood – is usually used for pK^+ as the influence of the above variables is negligible.

The measurement is performed at 37 °C. Conversion to other temperatures is possible. Calibration is performed with 2 calibration gases of exactly known CO_2 concentration. The pCO_2 of the calibration gases is calculated as follows:

$$pCO_2 = (p - pH_2O) \, 10^{-2} \times CO_2$$

pCO_2: CO_2 partial pressure (mm Hg)
p: atmospheric pressure according to current barometer reading (mm Hg)
pH_2O: saturated steam pressure of water (mm Hg) at 37 °C: 47 mm Hg (6.275 kPa)
CO_2:   volume fraction of CO_2 in the gas mixture in percent
  Reference temperature: 37 °C

Portable blood gas analyzers use CO_2 equilibrated buffer solutions for calibration.

Total CO₂ (Plasma)

The total CO_2 comprises dissolved CO_2, H_2CO_3, CO_3^{2-}, protein carbamates and ion pairs such as $NaHCO_3$, $CaHCO_3^+$ and $NaCO_3^-$ [126].

A reference method for determination of total CO_2 has been published [20]. It is an extraction method in which the CO_2 is released by addition of lactic acid and transferred to a vessel in which the CO_2 concentration is determined by titration with $BaCl_2/NaOH$:

$$CO_2 + H_2O + Ba^{2+} \rightarrow BaCO_3 + 2H^+$$
$$2H^+ + 2OH^- \rightarrow 2H_2O$$

The measured total CO_2 concentration is about 5% higher than the theoretically expected concentration [78]. The total CO_2 can be determined routinely by adding sulfuric acid to release the CO_2 and measuring the concentration with a pH dependent dye.

The total CO_2 concentration in the plasma is not usually measured but is calculated as part of the blood gas analysis using Formula 1:

$$CO_2 = \alpha_{CO_2} \times pCO_2 \times (1 + 10^{pH - pK^+})$$

α_{CO_2}: molar solubility coefficient for CO_2 (0.0307 mmol/(L × mm Hg) or 0.230 mmol/(L × kPa))
pCO_2: CO_2 partial pressure (mm Hg or kPa)
pK^+: apparent pK of CO_2 in blood (6.095)

or using Formula 2:

$$CO_2 = \alpha_{CO_2} \times pCO_2 + HCO_3^-$$

α_{CO_2}: molar solubility coefficient for CO_2 (mmol/L)
HCO_3^-: bicarbonate concentration in blood (mmol/L)
pCO_2: CO_2 partial pressure (mm Hg or kPa)

Actual Bicarbonate (Plasma)

The actual bicarbonate is the difference between total CO_2 on the one hand and dissolved CO_2 and H_2CO_3 on the other at the actual pCO_2, the actual hemoglobin oxygen saturation and the actual pH. There is no reference method available.

In routine practice the actual bicarbonate is seldom measured, use of appropriate ion selective electrodes has not yet become established. The actual bicarbonate is usually calculated as part of the blood gas analysis:

$$\log HCO_3^- = pH - pK^+ + \log pCO_2 + \log \alpha_{CO_2}$$

HCO_3^-: bicarbonate concentration in blood (mmol/L)
pK^+: apparent pK of CO_2 in blood (6.095)
pCO_2: CO_2 partial pressure (mm Hg or kPa)
α_{CO_2}: molar solubility coefficient for CO_2

Standard Bicarbonate (Plasma)

The standard bicarbonate is the plasma bicarbonate concentration after correction to pCO_2 40 mm Hg (5.33 kPa) at 37 °C and 100% oxygen saturation of hemoglobin by equilibration of the whole blood with CO_2. Use of the standard bicarbonate thus rules out any influence of the actual pCO_2 on the bicarbonate concentration.

There is no reference method available. Routinely the standard bicarbonate concentration can be calculated as part of the blood gas analysis. In spite of the pCO_2 standardization, the standard bicarbonate is not entirely independent of the original pCO_2 of the sample and its hemoglobin concentration. Nowadays the standard bicarbonate has been largely abandoned in favor of the actual bicarbonate or the base excess.

Base Excess (BE)

The base excess of the extracellular fluid is the base concentration measured after titration to pH 7.40 at a pCO_2 of 40 mm Hg (5.33 kPa) a temperature of 37 °C and actual oxygen saturation. For routine measurement a sample can be used which consists of one part (by vol.) of blood and two parts of the corresponding plasma. A reference method has not yet been defined. The base excess of the extracellular fluid is a measure of the non-respiratory component of an acid-base disorder. Usually the base excess of the extracellular fluid is not measured but calculated using the measurements obtained in the blood gas analysis.

$$BE_{ecf} = [HCO_3^-] - [HCO_3^-]^* + \beta_{ecf}(pH - pH^*)$$

HCO_3^-: actual bicarbonate (mmol/L)
$HCO_3^-{}^*$: 24.2 (mmol/L)
β_{ecf}: apparent buffer concentration of non-bicarbonate buffer in the extracellular fluid, 14.8 (mmol/L)
pH*: 7.40

The value used in the formula for β_{ecf} is not strictly constant but is dependent on the protein, phosphate and hemoglobin concentrations and on the volume of interstitial fluid. However, the deviations from the given value are so small that they are usually of no medical relevance.

The base excess of the blood can be determined by titrating blood to a plasma pH of 7.40 at a pCO_2 of 40 mm Hg (5.33 kPa), a temperature of 37 °C and the actual oxygen saturation. In routine testing the base excess of blood (BE_b) is calculated as part of the blood gas analysis.

$$BE_b = (1 - 0.014 \times C_{Hb})\left[C_{HCO_3^-} - C^*_{HCO_3^-} + (1.43 C_{Hb} + 7.7)(pH - pH^*)\right]$$

C_{Hb} hemoglobin concentration (g/dL)
$C^*_{HCO_3^-}$:   24.8 (mmol/L)
pH^*:   7.40

While the base excess of the extracellular fluid is not subject to any respiratory influences, this is not true of the base excess of blood (example: acute, uncompensated respiratory disorder). Therefore the base excess of the extracellular fluid is currently preferred as a measure of the solely non-respiratory component of the acid-base balance.

6.2 Blood Gases

pO_2 : O_2 Partial Pressure

The pO_2 is measured by amperometry. In amperometry the current is measured at a certain, constant potential of the working electrode in the region of the limiting current. The current is proportionate to the concentration of the depolarizing agent, e.g. O_2.

$$I = \frac{2 \times F \times D \times A}{\delta} \times c$$

I: current
A: surface of working electrode
2:    number of electrons exchanged in the electrode reaction
δ: thickness of the Nernst diffusion layer
F: Faraday's constant
c: concentration of the depolarizer in the electrode reaction
D: diffusion coefficient

pO_2 Determination

The pO_2 is determined using the Clark electrode. This consists of a platinum cathode to which a current of approx. −700 mV compared with the Ag/AgCl reference electrode is applied. The tip of the electrode is covered with a membrane which is permeable only for oxygen. Depending on the O_2 pressure of the test solution O_2 molecules penetrate the membrane and are reduced at the platinum cathode:

$$O_2 + 2H_2O + 4e^- \rightarrow 4OH^-$$

The electrodes are released by the reference electrode:

$$4Ag - 4e^- \rightarrow 4Ag^+$$

I.e. there is a current of 4 electrons for each molecule of oxygen. The OH^- ions combine with protons of the electrode electrolyte solution to form H_2O, which leads to a gradual pH shift in spite of the added buffer. The measurement is performed at 37 °C, conversion to a different temperature is unproblematic.

Gas/Blood Correction Factor
The platinum electrode has an intrinsic O_2 consumption; in addition O_2 dissolves in the membrane and sealing material. These phenomena affect only the pO_2 measurement of the blood sample; they do not affect the calibration gas mixture. For this reason the pO_2 in blood is 2–6% lower than in the gas mixture used for equilibration of the blood. This error is eliminated using a gas/blood correction factor.

Interferences
The pO_2 measurement is influenced to varying degrees by halothane.
No reference method for determination of pO_2 has been defined. Ideal reference material for verifying the accuracy of the results can be prepared by the individual laboratory by tonometry [19].

O_2 Saturation in Blood

The O_2 saturation (sO_2) gives the extent to which the hemoglobin is saturated with O_2. In this context hemoglobin is taken to mean only oxyhemoglobin (HbO_2) and deoxyhemoglobin (Hb).

$$sO_2 = \frac{HbO_2}{HbO_2 + Hb}$$

In the conventional calculation of the O_2 saturation the influence of dyshemoglobins (e.g. methemoglobin, CO-hemoglobin), abnormal hemoglobins (e.g. thalassemia), glycated hemoglobin and 2,3-diphosphoglycerate are not taken into account. Therefore in patients with increased concentrations of dyshemoglobins, for example, the O_2 saturation should be measured and not calculated.

O_2 Hemoglobin Fraction in Blood

The O_2 hemoglobin fraction ($FHbO_2$) gives the fraction of the total hemoglobin consisting of oxygen-carrying hemoglobin. In this context total hemoglobin is taken to mean oxyhemoglobin, deoxyhemoglobin and the dyshemoglobins (DysHb).

$$FHbO_2 = \frac{HbO_2}{HbO_2 + Hb + DysHb}$$

The O_2 hemoglobin fraction is determined by spectrophotometry.

Total O_2 Concentration in Blood (O_2 Content)

The total O_2 concentration (C_{tO2}) of the blood comprises the oxygen that is dissolved in the blood and the oxygen that is bound to hemoglobin (C_{HbO2}).

$$C_{tO2} = \alpha_{O2} pO_2 + c_{HbO2}$$
$$\alpha_{O2} : 10.5 \times 10^{-3} \, mmol \times L^{-1} \times kPa^{-1}$$

C_{HbO2} can be measured by spectrophotometry or calculated on the basis of the O_2 saturation (SO_2):

$$C_{HbO2} - C_{Hb} \times SO_2$$

CHb refers to the concentration of deoxy- and oxyhemoglobin (in mmol Hb(Fe)/L).

Tonometry

Protein-containing aqueous electrolyte solutions are available commercially for verifying the accuracy of pO_2 and pCO_2 measurement by a blood gas analyzer. Blood samples are not available as control samples as they are not sufficiently stable. However, the analytical chemist can use tonometry to prepare his own control material using blood from healthy subjects, for example. This eliminates the problem of matrix differences between patient samples and controls and also provides target values which can be classed as reference method values. Tonometry must be performed with highly pure gas mixtures of CO_2, O_2, and N_2, the exact composition of which must be specified by the manufacturer. The blood sample is placed in the tonometer chamber which has been adjusted to a temperature of 37 °C and the chamber is then shaken or rotated so that the blood forms a thin film on the inside wall. Then the gas mixture saturated with steam at 37 °C is passed through the chamber where it comes into contact with the blood. In this way a steady state between the liquid and the gas phase (equilibration) is quickly reached.

As the composition of the gas mixture is known the pCO_2 and pO_2 of the blood sample can be calculated:

1. $P_B = (P_{atm} - P_{H_2O}) \cdot V_f$

P_B: partial pressure of the gas in blood (kP)
P_{atm}: actual atmospheric pressure acc. to barometer (kP)
P_{H_2O}: steam pressure at 37 °C (= 6.27 kP)
V_f: volume fraction of the respective gas in the gas mixture

Example: P_{atm}: 102.91 ka
V_f: 0.25
$P_B = (102.91 - 6.27) \cdot 0.25 = 24.16$ [kPa]

or:

2. $P_{BI} = \dfrac{(P_{at} - P_{aq}) \cdot G}{100}$

P_{BI}: partial pressure of the gas in blood (mm Hg)
P_{at}: actual atmospheric pressure acc. to barometer (mm Hg)
P_{aq}: steam pressure at 37 °C (=47 mm Hg)
G: volume fraction of the gas in the gas mixture in %

Example: P_{at}: 772 mm Hg
G: 25%
$P_{BI} = \dfrac{(772 - 47) \cdot 25}{100} = 181$ [mm Hg]

6.3 Acid-Base Balance in Urine

Preliminary Remarks

On a normal diet a healthy person eliminates approximately 40–80 mmol protons per day with the urine. Only a very small fraction of the H^+ ions is eliminated in free form. For the most part the protons are either bound to NH_3 or NH_4^+ or buffered, which is detected as titratable acidity. The titratable acidity is based largely on the phosphate buffer system: $HPO_4^{2-} + H^+ \leftrightarrow H_2PO_4^-$.

Other acids play practically no role at all on account of their low pK and their usually lower concentrations compared with phosphate. Bicarbonate is only found – in amounts increasing with increasing pH – in urine with a pH above 6.5. That is, at a pH below 6.5 the net acid content of the urine is:

Ammonium (mmol/24 h) + titratable acidity (mmol/24 h)

At a pH above 6.5 the net acid content of the urine is:

Ammonium (mmol/24 h) + titratable acidity (mmol/24 h)
 – bicarbonate (mmol/24 h)

pH

In this context a finely graduated indicator paper is sufficient for determination of urine pH. Measurement with a pH electrode gives a reference interval of pH 5.53–6.97 for healthy individuals eating a normal diet. The lowest urinary pH is 4.5. In the context of the acid-base balance the urinary pH is determined to estimate whether bicarbonate needs to be taken into account in determination of the net acid content.

Ammonium

The ammonium concentration in urine can be determined by the Berthelot reaction, for example. Ammonia reacts with phenol and hypochlorite to form a blue dye.

Titration Acidity

At a pH below 7.4 the urine is titrated with sodium hydroxide solution to pH 7.4. The titration acidity is calculated on the basis of the amount of sodium hydroxide solution used and given in mmol/24 h. At a pH above 7.4 the urine is titrated with hydrochloric acid to pH 7.4 and the titration acidity, determined as described above, is negative.

Bicarbonate

At a urinary pH above 6.5 the bicarbonate concentration in 24h-urine must be determined for calculation of the net acid excretion. For this purpose CO_2 is released from the bicarbonate by addition of a strong acid and determined titrimetrically, for example.

7. Quality Assurance

Introduction

In Germany the quality assurance procedures for quantitative laboratory assays are described in the guidelines of the German Bundesärztekammer (1988, 1990, 1991) [3, 4, 5]. On 1 January 2002 new guidelines came into force which cover considerably more parameters than the old guidelines [1] and redefine the rules for performance of quality assurance.

The maximum allowable imprecision given in the guidelines is based on medical usefulness so that the analytical variation is small compared with the biological variation. This is achieved if one twelfth of the reference interval is used as allowable standard deviation for between-series imprecision [90]. The aim of these specifications is to ensure that the number of misclassifications (e.g. false reporting of pathologically elevated values) remains below a small, specifiable figure.

To determine the inaccuracy or bias, the arithmetic mean of the values obtained by measuring in control samples is compared with the target value for the parameter in question. For external quality control the type of target value is specified and should be a reference method value where possible.

Reference method values are determined using reference methods. In the methods hierarchy the reference method ranks above all corresponding routine methods for this parameter. In some cases the reference methods use procedures which are not suitable for routine testing, e.g. isotope dilution mass spectrometry, while in other cases more commonly used methods such as flame atomic emission spectrometry are employed. However, even in these cases the required reagent purity and precision and accuracy of the volumetric instruments as well as the metrological standards and the time required mean that reference methods are unsuitable for routine testing. Reference methods have been published for all electrolytes discussed here [16, 68, 95, 114, 115, 116, 117]. The instructions for performance of the methods are always very precise and detailed and must be adhered to exactly.

The so-called reference method value model has a number of advantages compared with the model based on method-dependent assigned values which was used in the past, namely

1. The accuracy of the routine method is assessed objectively.
2. Comparability of the analytical results is improved when the reference method value is the point of reference for all methods.
3. Inaccurate methods are identified and can be optimized.
4. New assay methods can be introduced at any time as there is no need to wait for the determination and publication of method-dependent assigned values.

Internal Quality Assurance

In every analytical series at least one accuracy control sample, i.e. a control sample with a target value (reference method value (preferred) or assigned value) should be measured. Control samples without target values, which were used in the past only for verification of precision, are no longer required by the new guidelines. From analytical series to analytical series control samples are to be used in different concentration ranges if at least 15 analytical series are performed within 3 months. The general rule is that every measuring instrument used for quantitative determination of a parameter must be included in the internal (and external) quality assurance for this parameter. In a kind of "run-up period" the intralaboratory error limits for individual control sample measurements are determined. For this purpose 20 control sample results from 20 consecutive shifts are selected according to the same pattern (e.g. always the first) for each control sample used. The arithmetic mean, the coefficient of variation and the standard deviation are calculated from the results obtained. These parameters may be used to set up the control card for the respective control sample provided that

1. the deviation of the arithmetic mean from the target value is smaller than or equal to the value given in the relevant annex of the guidelines ("maximum allowable inaccuracy");
2. the coefficient of variation or the standard deviation is smaller than or equal to the value given in the relevant annex of the guidelines ("maximum allowable imprecision").

To safeguard the quality during the "run-up period", during this phase the deviation of the result of a control sample from the target value must not exceed the "maximum allowable deviation of a single value" specified in the guidelines.

The following data are entered on the control card:

1. The arithmetic mean calculated by the laboratory for the analytical series on the respective instrument.
2. The mean ($\bar{x}$) ± 3 standard deviations as intralaboratory error limits.

After completion of the "run-up period" each individual control sample measurement is evaluated on the basis of the control card. The values should always be within the intralaboratory error limits ($\bar{x}$ ± 3 s) and should not exceed the maximum allowable deviation of a single value from the target value. Otherwise the responsible officer must decide whether it is necessary to completely or partially repeat the analytical series.

At the end of each control cycle, usually at the end of the month,

1. the between-series imprecision is checked by calculation of the appropriate standard deviation and coefficient of variation; these must not exceed the values given for the "maximum allowable imprecision". Otherwise the cause must be sought and eliminated;
2. the systematic error (inaccuracy) is checked by calculating the arithmetic mean of the results for the respective control material and comparing this with the target value. The systematic error must not exceed the "maximum allowable inaccuracy" given in the annexes of the guideline. Otherwise the cause must be sought and eliminated.

Rules for Special Cases

Quality Assurance in the Case of Low Test Frequencies

If fewer than 15 analytical series in 3 months are performed for a particular parameter on a particular instrument two control samples with significantly different analytical concentrations must be analyzed in each series. Neither result should exceed the "maximum allowable error" specified for a single measurement. Otherwise the cause must be determined and eliminated. Determination of the imprecision and the systematic error is not required. It is not necessary to keep a control card. However participation in proficiency tests is required.

Quality Assurance of Point-of-Care Testing in Doctors' Practices and Health Centers

The instruments must be checked with a physical and/or electronic standard at least once a day on the days when the instrument is used. A control sample must be measured once a week. The difference between the measured value and the target value is evaluated on the basis of the "maximum allowable deviation of a single value" given in the annexes to the guideline. If the specified value is exceeded the cause must be sought and eliminated.

Control samples with distinctly different analyte concentrations are to be used alternately. Participation in interlaboratory comparisons (proficiency testing) is not necessary.

The above regulations apply only to instruments which are intended or used exclusively for measurement of single samples.

Quality Assurance for Near-Patient Testing in Hospitals and Other Facilities with a Central Laboratory

The instruments concerned must be checked with a physical and/or electronic standard at least once a day on the days when the instrument is used. A control sample must be measured once a week. The difference between the respective measured value and the target value must not exceed the specifications of the guideline with regard to the "maximum allowable deviation of a single value", otherwise the cause must be sought and eliminated. Control samples with distinctly different analyte concentrations are to be used alternately.

In principle, participation in interlaboratory comparisons (proficiency testing) is required. However, participation is not necessary if the internal quality assurance procedures described above are performed under the responsibility of the central laboratory.

All regulations apply only to instruments which are intended or used exclusively for measurement of single samples.

External Quality Assurance (Proficiency Testing)

The proficiency testing is performed by institutions which are appointed to do so by the German *Bundesärztekammer*. For the parameters named in the annexes to the guideline participation in one proficiency testing survey per quarter is mandatory. The values determined by the participants are assessed on the basis of the specified "maximum allowable deviation of a single value". If the values measured for all (usually two) samples of the proficiency test for a particular parameter are below the specified limit the participant is issued a certificate for this parameter. This certificate is valid for 6 months.

Sodium (Serum, Plasma, Blood)

<u>Precision and Accuracy</u>
The relative standard deviation of the day-to-day imprecision must be below 1.5% [1, 67]. As far as possible, the accuracy of the methods for determination of the sodium concentration in serum should be checked using reference method values. The maximum allowable inaccuracy is ± 2%, the maximum allowable deviation of a single measurement ± 5%. These specifications also apply to "direct" determinations in undiluted serum or blood by ISE (ionized sodium). They can be met if the control samples correspond to "normal" serum samples (see 5.1) and the ionized sodium is measured according to IFCC recommendations [17].

<u>Plausibility control</u>
The sodium concentration in serum is usually closely correlated with the chloride concentration. Sodium concentrations below 122 mmol/L and above 160 mmol/L seldom occur and should be checked [119].

Chloride (Serum, Plasma)

<u>Precision and accuracy</u>
The relative standard deviation of the day-to-day imprecision must be below 2%. Where possible, the accuracy should be checked by comparing the measured values with the reference method values of the cor-

responding control serum [67]. The maximum allowable inaccuracy is ± 4%, the maximum allowable deviation of a single measurement 8%.

Plausibility control
Values below 83 mmol/L and above 126 mmol/L seldom occur and should be checked. Usually the sodium and chloride concentrations change in the same direction. Calculation of the anion gap can also help to detect errors.

Osmolality (Serum, Plasma)

Precision and Accuracy
The guidelines [1, 3] do not cover the determination of osmolality. Under routine conditions the average coefficient of variation for day-to-day imprecision over a period of three months was 1.6%. The accuracy was checked using method dependent assigned values.

Plausibility Control
Serum values below 255 mmol/kg and above 393 mmol/kg are anomalous and should be checked.

Potassium (Serum, Plasma, Blood)

Precision and Accuracy
The relative standard deviation of the day-to-day imprecision must be below 2.7% [1]. If possible, the accuracy of the potassium concentration determination should be checked by comparison with control sera for which reference method values are given [67]. The maximum allowable inaccuracy is ± 3.7%, the maximum allowable deviation of a single measurement ± 9.1%. These specifications also apply to "direct" determinations in undiluted serum or blood by ISE (ionized potassium, cf. sodium).

Plausibility Control
Potassium concentrations below 6.5 mmol/L and above 2.6 mmol/L seldom occur and should therefore be checked. Values above 10 mmol/L

are considered incompatible with life. If such values are obtained, the possibility of errors during sample collection (potassium-containing infusion solution?) and sample storage must be considered.

Magnesium (Serum, Plasma)

<u>Precision and Accuracy</u>
At concentrations > 0.8 mmol/L the relative standard deviation of day-to-day imprecision must not exceed 4%. At concentrations < 0.8 mmol/L the standard deviation must not exceed 0.032 mmol [1]. If possible, the accuracy should be checked on the basis of the reference method values of the control sera [66].

The maximum allowable inaccuracy (figures in brackets are for concentrations < 0.8 mmol/L) is ± 7% (± 0.056 mmol/L) and the maximum allowable deviation of a single value ± 15% (± 0.12 mmol/L).

The "guidelines" do not give any specifications with regard to ionized magnesium. For the time being the specifications for total magnesium should be used. The accuracy must be checked with method dependent assigned values as a reference method for ionized magnesium is not yet available.

<u>Plausibility Control</u>
Total magnesium concentrations below 0.38 mmol/L and above 2.07 mmol/L seldom occur and should be checked.

Calcium, Total (Serum, Plasma)

<u>Precision and Accuracy</u>
The day-to-day imprecision must be less than 3% [1]. The accuracy should be checked by comparing the results obtained by analysis of control sera with the reference method values [66]. The maximum inaccuracy should be ± 5%, the maximum deviation of a single value ± 11%.

<u>Plausibility Control</u>
Calcium concentrations below 1.5 mmol/L and above 3.4 mmol/L seldom occur and should be checked. In most disorders (except rickets) the product of calcium and phosphate in serum is largely constant.

Calcium, Ionized (Blood)

Precision and Accuracy
The maximum allowable imprecision on determination of ionized calcium in blood is 5% (± 0.05 mmol/L) [1]. The maximum allowable inaccuracy (figures in brackets are for concentrations < 1 mmol/L) is ± 5% (± 0.05 mmol/L) and the maximum allowable deviation of a single value ± 15% (± 0.15 mmol/L).
A reference method is available [18, 27] but is not yet widely used for determination of reference method values.

Plausibility Control
Various formulae [97] can be used to calculate the "ionized calcium" fraction from the total calcium and albumin. However, there is often only a rough agreement between the calculated and measured concentrations. Values below 0.85 mmol/L and above 1.70 mmol/L seldom occur and should be checked.

Phosphate

Precision and Accuracy
The day-to-day coefficient of variation for imprecision must not exceed 5% [1]. If possible the accuracy should be checked on the basis of reference method values [95]. The maximum allowable inaccuracy is ± 8%, the maximum allowable deviation of a single measurement 18%.

Plausibility Control
There is often a negative correlation between calcium and phosphate in serum (exception: rickets); the product of the two parameters therefore only fluctuates within narrow limits. Serum concentrations below 0.30 mmol/L and above 3.33 mmol/L should be checked.

Electrolytes in Urine
The requirements of the Bundesärztekammer guideline [1] regarding the reliability of electrolyte determinations in urine are summarized in Tab. 57.

Table 57. Electrolytes in urine [1]

Electrolyte	Target Value	Maximum Allowable Imprecision	Maximum Allowable Inaccuracy	Maximum Allowable Deviation of a Single Value	Measuring Range
Sodium	RMV	3% 2.4 mmol/L	5% 4.0 mmol/L	11% 8.8 mmol/L	≥ 80 mmol/L < 80 mmol/L
Chloride	RMV	4%	6%	14%	
Potassium	RMV	5%	7%	17%	
Magnesium	RMV	6% 0.06 mmol/L	8% 0.08 mmol/L	20% 0.20 mmol/L	≥ 1 mmol/L < 1 mmol/L
Calcium	AV	5% 0.1 mmol/L	5% 0.1 mmol/L	15% 0.3 mmol/L	≥ 2 mmol/L < 2 mmol/L

RMV: reference method value, AV: method-dependent assigned value.

pH (Blood)

Precision and Accuracy
The maximum allowable imprecision is ± 0.02 [1]. In the comparison with the target value the maximum allowable inaccuracy is ± 0.02, the maximum allowable deviation of a single value 0.06.

Unfortunately the only control material currently available for quality assurance has properties that usually differ greatly from the actual test material, namely blood. Reference method values are available [77].

Plausibility Control
On account of the close relationship between pH, pCO_2 and actual bicarbonate it is appropriate to look at all three parameters together to see whether the results are plausible. A pH below 7.21 or above 7.57 seldom occurs and should therefore be checked.

pCO_2 (Blood)

Precision and Accuracy
The guidelines allow a maximum imprecision of 4% [1]. The maximum allowable inaccuracy on comparison with the assigned values is ± 4%, the maximum allowable deviation of a single value 12%.

The commercially available control material differs greatly from the actual test material (see pH). However, laboratories can use tonometry to produce their own reference material (see page 91–92) [19].

Plausibility Control
As described for pH, the plausibility can first be checked by looking at the three parameters pH, pCO_2 and actual bicarbonate together. Extreme values which require checking are a pCO_2 of below 25 mm Hg (3.33 kPa) or above 70 mm Hg (9.31 kPa).

Calculated Parameters
Base excess and actual bicarbonate are calculated parameters whose precision and accuracy depend on the precision and accuracy of the measurements from which they are calculated. However, it is important to point out that – as in the case of other parameters of blood gas analysis and acid-base balance, too – different computer programs from different manufacturers do not perform the calculations in an identical manner. This can be the cause of discrepant results.

pO_2 (Blood)

Precision and Accuracy
According to the guidelines [1] the maximum allowable imprecision is 4% (4 mm Hg). In the comparison with the target values the maximum allowable inaccuracy is ± 4% (4 mm Hg), the maximum allowable deviation of a single value 12% (12 mm Hg). The numbers in brackets are for pO_2 < 100 mm Hg.

Commercial control material is available for quality control. However, this material differs considerably from the actual test material. Therefore it is in principle preferable for laboratories to prepare their own reference material using tonometry (see page 91–92) [19].

Plausibility Control

Extreme values are: pO_2 < 50 mm Hg

Mechanical ventilation with 100% oxygen can produce pO_2 values above 500 mm Hg.

Appendix

Table 58. Reference intervals: Electrolyte concentrations in serum (S)

S-Calcium, total	2.15–2.55 mmol/L	[41]
S-Calcium, ionized (pH 7.4)	1.17–1.29 mmol/L	[96, 97, 98]
S-Chloride	98–106 mmol/l.	[109]
S-Magnesium, total	0.65–1.05 mmol/L	[2, 124]
S-Magnesium, ionized (pH 7.4)	0.53–0.67 mmol/L	[6]
S-Osmolality	275–295 mosmol/kg	[55]
S-Phosphate, inorganic	0.87–1.45 mmol/L	[109]
S-Potassium	3.5–5.1 mmol/L	[109]
S-Sodium	135–145 mmol/L	[59]

The reference intervals are to be considered as guidelines and apply to serum samples from adults [100, 101, 102].

Table 59. Reference intervals: Electrolyte excretion in 24 h-collected urine (U)

U-Calcium	2.5–7.5 mmol/d	[109]
U-Chloride	110–250 mmol/d	[109]
U-Magnesium	3.0–5.0 mmol/d	[109]
U-Osmolality	300–900 mosmol/kg	[109]
U-Phosphate, inorganic	12.9–42.0 mmol/d	[109]
U-Potassium	25–125 mmol/d	[109]
U-Sodium	27–287 mmol/d	[109]

The reference intervals are to be considered as guidelines and apply to urine samples from adults.

Table 60. Reference intervals: Electrolyte concentrations in 1st morning urine (U)

U-Calcium	1.3–8.9 mmol/L	[60]
U-Chloride	46–168 mmol/L	[60]
U-Magnesium	1.6–5.7 mmol/L	[60]
U-Phosphate, inorganic	13–44 mmol/L.	[60]
U-Potassium	20–80 mmol/L	[60]
U-Sodium	54–190 in mmol/L	[60]

The reference intervals are to be considered as guidelines and apply to urine samples from adults.

Table 61. Reference intervals: Urinary indices

	Clearance* [mL/min]	Fractional Excretion [%]	Tubular Resorption [%]
Chloride	0.7–2.1	0.7–1.7	98–99
Osmolality	2–4	< 3.5	
Phosphate	5.4–16.2		82–90
Potassium	6–20	5–17	83–95
Sodium	0.7–0.8	1–3	97–99
Water, free	− 0.4 to − 2.5	1–2	98–99

*With reference to 1.73 m² body surface area.

Table 62. Changes in sodium concentration in serum (mmol/L) related to protein and lipid concentrations. Modified from Levy [74]

	Protein [g/L]						
Lipids [g/L]	45	55	65	75	85	95	105
2.5	149.2	148.1	146.9	145.8	144.7	143.5	142.4
7.5	148.4	147.3	146.1	145.0	143.9	142.7	141.6
12.5	147.6	146.5	145.3	144.2	143.1	141.9	140.8
17.5	146.8	145.7	144.5	143.4	142.3	141.1	140.0
22.5	146.0	144.9	143.7	142.6	141.5	140.3	139.2
27.5	145.2	144.1	142.9	141.8	140.6	139.5	138.4
32.5	144.4	143.3	142.1	141.0	139.8	138.7	137.6
37.5	143.6	142.5	141.3	140.2	139.0	137.9	136.8
42.5	142.8	141.7	140.5	139.4	138.2	137.1	136.0
47.5	142.0	140.9	139.7	138.6	137.4	136.3	135.2
52.5	141.2	140.0	138.9	137.8	136.6	135.5	134.3
57.5	140.4	139.2	138.1	137.0	135.8	134.7	133.5
	Protein [g/L]						
Lipids [g/L]	115	125	135	145	155	165	175
2.5	141.2	140.1	139.0	137.8	136.7	135.5	134.4
7.5	140.1	139.3	138.2	137.0	135.9	134.7	133.6
12.5	139.6	138.5	137.4	136.2	135.1	133.9	132.8
17.5	138.8	137.7	136.6	135.4	134.3	133.1	132.0
22.5	138.0	136.9	135.8	134.6	133.5	132.2	131.2
27.5	137.2	136.1	134.9	133.8	132.7	131.5	130.4
32.5	136.4	135.3	134.1	133.0	131.9	130.7	129.6
37.5	135.6	134.5	133.3	132.2	131.1	129.9	128.8
42.5	134.8	133.7	132.5	131.4	130.3	129.1	128.0
47.5	134.0	132.9	131.7	130.6	129.5	128.3	127.2
52.5	133.2	132.1	130.9	129.8	128.6	127.5	126.4
57.5	132.4	131.3	130.1	129.0	127.8	126.7	125.6

Table 63. Conversion of concentration data

Analyte	Unit	Factor	Unit
Calcium	mmol/L	2	mEq/L
	mmol/L	4.008	mg/dL
Chloride	mmol/L	1	mEq/L
	mmol/L	3.545	mg/dL
Magnesium	mmol/L	2	mEq/L
	mmol/L	2.431	mg/dL
Phosphate (PO_4^{3-})	mmol/L	9.497	mg/dL
Phosphorus	mmol/L	3.097	mg/dL
Potassium	mmol/L	1	mEq/L
	mmol/L	3.910	mg/dL
Sodium	mmol/L	1	mEq/L
	mmol/L	2.299	mg/dL

In order to calculate the mass concentration (mg/dL), for example, the corresponding figure for the substance concentration (mmol/L) must be multiplied by the given factor.

The substance concentration (mmol/L) is calculated, for example, from the mass concentration (mg/dL) by dividing the corresponding figure by the given factor.

Table 64. Conversion of pH data

		Factor		H^+ Concentration
pH	x	$10^{-pH} \times 10^9$	→	nmol/L

Table 65. Conversion of pressure units

Pressure		Factor		Pressure
mm Hg	x	0.133	→	kPa
kPa	x	7.5	→	mm Hg
cm H_2O	x	0.098	→	kPa
kPa	x	10.2	→	cm H_2O
mm Hg	x	1	→	Torr
bar	x	100	→	kPa

Table 66. Reference intervals: Acid-base balance and blood gases (adults)

Arterial blood [25, 86]	
pH	7.37–7.45 (37 °C, sea level)
H$^+$ Concentration	42.7–35.5 mmol/L
pCO_2	
Males	4.66–6.12 kPa
	35.0–46.0 mm Hg
Females	4.26–5.72 kPa
	32.0–43.0 mm Hg
Actual bicarbonate (plasma)	21–26 mmol/L
Standard bicarbonate	21–26 mmol/L
Total CO_2 (plasma)	23–28 mmol/L
Base excess of extracellular fluid	−2–+3 mmol/L
pO_2 (arterial blood)		

Age (years)	kPa	mm Hg
20–29	11.2–13.9	84–104
30–39	10.8–13.5	81–101
40–49	10.4–13.1	78–94
50–59	9.9–12.5	74–91
60–69	9.5–12.1	71–91
O_2 Saturation	0.950–0.985	
HbO_2 Fraction	0.94–0.98	
Total O_2 concentration	180–230 ml/L	
	8.0–10.3 mmol/L	

Urine [86]	
pH	5.53–6.97
Ammonia	20–50 mmol/24 h
Titration acidity	10–40 mmol/24 h
Net acid	40–80 mmol/24 h

Table 67. Reference intervals: Lymph [113]

Calcium	1.7–3.0 mmol/L
Chloride	85–130 mmol/L
Potassium	3.8–5.0 mmol/L
Sodium	104–108 mmol/L
pH	7.4–7.8

Table 68. Reference intervals: Gastric juice [28]

Calcium	1.0–2.3 mmol/L
Chloride	7.8–159 mmol/L
Magnesium	0.25–1.50 mmol/L
Phosphate	0.19–5.8 mmol/L
Potassium	6.5–16.5 mmol/L
Sodium	19–70 mmol/L

Table 69. Reference intervals: Bile, colorless [112]

Chloride	108–138 mmol/L
Magnesium	< 0.2 mmol/L
Phosphate	< 0.6 mmol/L
Potassium	3.9–5.7 mmol/L
Sodium	144–156 mmol/L
pH	7.09–8.01

Table 70. Reference intervals: Saliva [40, 47]

Calcium	1.2–2.8 mmol/L
Chloride	5–40 mmol/L
Magnesium	0.08–0.53 mmol/L
Phosphate	0.45–12.5 mmol/L
Potassium	10–36 mmol/L
Sodium	2–21 mmol/L
pH	5.1–7.3

Table 71. Reference intervals: Sweat [9, 40, 49]

Chloride	approx. 30 mmol/L
Potassium	approx. 7.5 mmol/L
pH	6.0–6.6
Pilocarpine iontophoresis	
Chloride	< 50 mmol/L
Potassium	< 40 mmol/L

Table 72. Reference intervals: Liquor [58, 59]

Calcium	1.05–1.35 mmol/L
Potassium	3 mmol/L
Sodium	141 mmol/L

Table 73. Reference intervals: Nasal secretion [33, 59]

Calcium	1.00–1.75 mmol/L
Potassium	17 mmol/L
Sodium	90–148 mmol/L

Table 74. Reference interval: Amniotic fluid, potassium [46]

up to 20th week of gestation	3.3–4.6 mmol/L
> 20th week of gestation	2.9–5.6 mmol/L

References

1. Richtlinie der Bundesärztekammer zur Qualitätssicherung quantitativer laboratoriumsmedizinischer Untersuchungen. Dt Ärztebl 2001; 98: A2747–59.
2. Food and Nutrition Board, Institute of Medicine. Dietary reference intakes for calcium, phosporus, magnesium, vitamin D and fluoride. Washington, DC, National Academy Press 1997.
3. Richtlinien der Bundesärztekammer zur Qualitässicherung in medizinischen Laboratorien. Frist für Übergangsregelungen verlängert. Dt Arztebl 1991; 88: B222–3.
4. Qualitätssicherung der quantitativen Bestimmungen im Laboratorium. Neue Richtlinien der Bundesärztekammer. Dt Arztebl 1988; 85: B517–32.
5. Übergangsregelungen zu den Richtlinien der Bundesärztekammer zur Qualitätssicherung in medizinischen Laboratorien aufgrund des Beschlusses des Vorstandes der Bundesärztekammer vom 15. Dezember 1989. Dt Arztebl 1990; 87: B363–5.
6. Altura BT, Shirey TL, Young CC, Dell'Orfano K, Altura BM. Characterization and studies of a new ion selective electrode for free extracellular magnesium ions in whole blood, plasma and serum. In: D'Orazio P, Burritt MF, Sena SF, eds. Electrolytes, blood gases and other critical analytes: The patient, the measurement and the government. Vol. 14. Madison, Omnipress 1992.
7. Bablok W, Passing H, Bender R, Schneider B. A general regression procedure for method transformation. J Clin Chem Clin Biochem 1988; 26: 783–90.
8. Bates RG, ed. Determination of pH. Theory and practice. 3rd edition. New York, John Wiley & Sons 1973: 59–104.
9. Barnes GL, Vaelioja L, McShane S. Sweat testing by capillary collection and osmometry: suitability of the Wescor Macroduct system for screening suspected cystic fibrosis patients. Aust Paediatr J 1988; 24: 191–3.
10. Battle DC, Nizon M, Cohen E. The use of the urinary gap in the diagnosis of hyperchloremic metabolic acidosis. New Engl J Med 1988; 318: 594–9.

11. Berry MN, Mazzachi RD, Peake MJ. Enzymatic spectrophotometric determination of sodium, potassium, and chloride ions in serum or urine. Wien Klin Wochenschr 1992; 104 (Suppl. 192): 5–11.
12. Berry MN, Mazzachi RD, Pejakovic M, Peake MJ. Enzymatic determination of sodium in serum. Clin Chem 1988; 34: 2295–8.
13. Berry MN, Mazzachi RD, Pejakovic M, Peake MJ. Enzymatic determination of potassium in serum. Clin Chem 1989; 35: 817–20.
14. Boening D, Maassen N. Blood osmolality in vitro: Dependence on pCO_2, lactic acid concentration, and O_2 saturation. J Appl Physiol 1983; 54: 118–22.
15. Boening D, Schweigart U, Kunze M. Diurnal variations of protein and electrolyte concentrations and of acid base status in plasma and red cells of normal man. Eur J Appl Physiol 1974; 32: 239–50.
16. Brown SS, Healy MJR, Kearns M. Report on the interlaboratory trial of the reference method for the determination of total calcium in serum. Part 1. J Clin Chem Clin Biochem 1981; 19: 395–412.
17. Burnett RW, Covington AK, Fogh-Andersen N, et al. Recommendations for measurement of conventions for reporting sodium and potassium by ion-selective electrodes in undiluted serum, plasma or whole blood. Clin Chem Lab Med 2000; 38: 1065–71.
18. Burnett RW, Christiansen TF, Covington AK, et al. IFCC recommended reference method for the determination of the substance concentration of ionized calcium in undiluted serum, plasma or whole blood. Clin Chem Lab Med 2000; 38: 1301–14.
19. Burnett RW, Covington AK, Maas AHJ, et al. IFCC method (1988) for tonometry of blood: Reference materials for pCO_2 and pO_2. Clin Chim Acta 1989; 185: 17–24.
20. Burnett RW, Covington AK, Fogh-Andersen N, et al. IFCC reference measurement procedure for substance concentration determination of total carbon dioxide in blood, plasma or serum. Clin Chem Lab Med 2001; 39: 283–9.
21. Burnett RW, Covington AK, Fogh-Andersen N, et al. Approved recommendations on whole blood sampling, transport and storage for simultaneous determination of pH, blood gases and electrolytes. Eur J Clin Chem Clin Biochem 1995; 33: 247–53.
22. Cammann K. Elektrochemische Untersuchungsverfahren. In: Naumer H, Heller W, eds. Untersuchungsmethoden in der Chemie. Stuttgart New York, Thieme 1986.

23. Chapoteau E, Gebauer CR, Chimenti MZ, et al. Principles and performance of the Technicon Chromolyte™ sodium method. Accuracy, interferences and multisite imprecision studies. Clin Chem 1990; 36: 1065–8.
24. Champion P, Pellet H, Grenier M. Spectronatromètre. L'analyse spectrale par la flamme. CN Acad Sci 1873; 76: 707–10.
25. Cohen MR, Feldman GM, Fernandez PC. The balance of acid, base and charge in health and disease. Kidney Int 1997; 52: 287–93.
26. Colombo JP, ed. Klinisch-chemische Urindiagnostik. Arbeitsgruppe Urin der Schweiz Ges f Klin Chem. Rotkreuz, Labo Life Verlagsgemeinschaft 1994.
27. Covington AK, Kelly PM, Maas AHJ. Reference method for the determination of ionized calcium in serum, plasma or whole blood: Experimental aspects. In: Maas AHJ, Buckley BM, Manzoni A, et al., eds. Methodology and Clinical Applications of ion-selective electrodes. Vol. 10. Utrecht, Elinkwijk Printing 1988.
28. Dennebaum R. Extravasale Körperflüssigkeiten. In Thomas L, ed. Labor und Diagnose. 6. Auflage; TH-Books, Frankfurt/Main 2005: 1812–3.
29. Duarte GC. Disorders in magnesium metabolism. In: Chang J, Gill JR, eds. Kidney electrolyte disorders. Livingstone, Churchill 1990: 290.
30. Durst RA, ed. Standard Reference Materials: Standardization of pH Measurements. NBS Special Publications 260–53. Washington DC, US Government Printing Office 1975: 1–35.
31. Dybkaer R, Solberg HE. Approved recommendation (1987) on the theory of reference values. Part 6: Presentation of observed values related to reference values. J Clin Chem Clin Biochem 1987; 25: 657–62.
32. Ehrhardt V, Appel W, Paschen L, et al. Evaluierung eines Xylidyl-Blau-Reagenz zur Bestimmung von Magnesium. Wien Klin Wochenschr 1992; 104 (Suppl. 192): 5–11.
33. Eichner H, Bebbehani AA, Hochstraßer K. Nasensekretdiagnostik – aktueller Stand. Normalwerte. Laryng Rhinol Otol 1983; 62: 561–5.
34. Elfin RJ, Robertson EA, Johnson E. Bromide interferes with determination of chloride by each of four methods. Clin Chem 1981; 27: 778–9.
35. Ehrhardt W, Neumann H, Schmidt LH, Wessig H. Säure-Basen-Gleichgewicht des Menschen. Dresden, Steinkopff 1976.

36. Farrell EC. Phosphorus. In: Pesce AJ, Kaplan LA, eds. Methods in clinical chemistry. St. Louis Washington Toronto, CV Mosby Company 1987: 881–4.
37. Farrell EC. Calcium. In: Pesce AJ, Kaplan LA, eds. Methods in clinical chemistry. St. Louis Washington Toronto, CV Mosby Company 1987: 865–9.
38. Fiedler H, Lieb L, Zimmermann K. Computerunterstützte Nierenfunktionsüberwachung in der Intensivmedizin. J Clin Chem Clin Biochem 1990; 28: 712–3.
39. Fossati P, Sirtoli M, Tarenghi G, Giachetti M, Berti G. Enzymatic assay of magnesium through glucokinase activation. Clin Chem 1989; 35: 2112–6.
40. Ganong WF, Auerswald F. Lehrbuch der medizinischen Physiologie. Berlin, Springer 2002.
41. Gosling P. Analytical reviews in clinical biochemistry: Calcium measurement. Ann Clin Biochem 1986; 23: 146.
42. Gindler EM, King JD. Regulation of the intracellular environment: Potassium and other salts. Am J Clin Pathol 1972; 58: 376–9.
43. Gray TA, Buckley BM, Sealey MM, Smith SCH, Tomlin P, McMaster P. Plasma ionized calcium monitoring during liver transplantation. Transplantation 1986; 41: 335–9.
44. Guder WG, Lang H, eds. Pathobiochemie und Funktionsdiagnostik der Niere. Berlin Heidelberg New York, Springer 1991.
45. Guder WG. Niere und ableitende Harnwege. In: Greiling H, Gressner AM, eds. Lehrbuch der klinischen Chemie und Pathobiochemie, 2. Aufl. Stuttgart New York, Schattauer 1990.
46. Guibaud S, Bonnet M, Dury A, Thoulon JM, Dumont M. Amniotic fluid or maternal urine? Lancet 1976; 1: 746
47. Haeckel R, Walker RF, Colic D. Reference ranges for mixed saliva collected from the literature. J Clin Chem Clin Biochem 1989; 27: 249–52.
48. Halperin ML, Goldstein MB. Fluid, Electrolyte and Acid Base Emergencies. Philadelphia, WB Saunders 1988.
49. Heinrich HG, Adler D, Jung K, Jaroß H, Rose W. Vergleichende Untersuchungen der Konzentration von β_2–Mikroglobulin, α_1-Mikroglobulin und Lysozym in Serum, Urin, Schweiß und Speichel bei gesunden und Nierenkranken. Klin Lab 1991; 37: 377–81.

50. Holden NE, Martin RL. Atomic weights of the elements. Pure Appl Chem 1981; 55: 1011–8.
51. Honold F, Honold B. Ionensensitive Elektroden. Basel Boston Berlin, Birkhäuser 1991.
52. Jansen H, Kerscher M, Town M. Evaluation of enzymatic methods for the determination of sodium and potassium on the Hitachi 717 and 704. Clin Lab 1991; 37: 303–4.
53. Kaplan SA, Yuceoglu AM, Strauss J. Chemical microanalysis: Analysis of capillary and venous blood. Pediatrics 1959; 24: 270–4.
54. Kayamori Y, Katayama Y. Enzymatic method for assaying calcium in serum and urine with porcine pancreatic amylase. Clin Chem 1994; 40: 781–4.
55. Keller H. Klinisch-chemische Labordiagnostik für die Praxis, Analyse, Befund, Interpretation. 2. Aufl., Stuttgart New York; Thieme 1991.
56. Kessler G, Wolfman M. An automated procedure for the simultaneous determination of calcium and phosphorus. Clin Chem 1964; 10: 686–703.
57. Kirchhoff GR, Bunsen RW. Alkali- und Erdalkali-Spektren. Ann Physik Chem 1991; 110: 161.
58. Kjeldsberg CR, Krieg AF. Cerebrospinal fluid and other body fluids. In: Henry JB, ed. Clinical diagnosis and management by laboratory methods. Philadelphia, Saunders 1984: 459–92.
59. Kortekangas AE. Funktion und Funktionsprüfung der Nase und der Nasennebenhöhlen. In: Link R, ed. Obere und untere Luftwege. Stuttgart, Thieme 1977.
60. Krieg M, Gunßer KJ, Steinhagen-Thiessen E, Becker H. Vergleichende Analytik im 24-h- und Morgenurin. J Clin Chem Clin Biochem 1986; 24: 863–9.
61. Külpmann WR. Influence of protein on the determination of sodium, potassium and chloride in serum by Ektachem DT 60 with the DTE module; evaluation with special attention to a possible protein error by flame atomic emission spectrometry and ionselective electrodes; proposals to their calibration. J Clin Chem Clin Biochem 1989; 27: 815–24.
62. Külpmann WR. Reference methods for the determination of sodium, potassium, pH and blood gases with ion-selective electrodes. Eur J Clin Chem Clin Biochem 1991; 29: 263–7.

63. Külpmann WR, Maibaum P, Sonntag O. Analyses with the KODAK-Ektachem. Accuracy control using reference method values and the influence of protein concentration. J Clin Chem Clin Biochem 1990; 28: 825–33.
64. Külpmann WR. Electrometric reference methodologies: Sodium and potassium, pH and blood gases. Fresenius' J Anal Chem 1990; 337: 8–9.
65. Külpmann WR. Determination of electrolytes in serum and serum water. Wien Klin Wochenschr 1992; 104 (Suppl. 192): 34–8.
66. Külpmann WR, Buchholz R, Dyrssen C, Ruschke D. A comparison of reference method values for calcium, lithium and magnesium with method-dependent assigned values. J Clin Chem Clin Biochem 1969; 27: 631–7.
67. Külpmann WR, Lagemann J, Sander R, Maibaum P. A comparison of reference method values for sodium, potassium and chloride with method-dependent assigned values. J Clin Chem Clin Biochem 1985; 23: 865–74.
68. Külpmann WR, Ruschke D, Büttner J, Paschen K. A candidate reference method for the determination of magnesium in serum. J Clin Chem Clin Biochem 1989; 27: 33–9.
69. Külpmann WR, Gerlach M. Relationship between ionized and total magnesium in serum. Scand J Clin Lab Invest 1996; 56 (Suppl. 224): 251–8.
70. Külpmann WR, Kallien T, Lewenstam A. Evaluation of an ion-selective electrode for the determination of "ionized" magnesium. In: D'Orazio P, Burritt MF, Sena SF, eds. Electrolytes, blood gases and other critical analytes: The patient, the measurement and the government. Madison, WI, Omnipress 1992: 188–211.
71. Külpmann WR, Rademacher E, Bornscheuer A. Ionized magnesium concentration during liver transplantation, resection of the liver, and cardiac surgery. J Clin Lab Invest 1996; 56 (Suppl. 224): 235–43.
72. Külpmann WR. Electrolytes: Determination in physiological samples. In: Townshend A, ed. Encyclopedia of analytical science. London, Academic Press 1995: 1003–12.
73. Külpmann WR, Höbbel T. International consensus of the standardization of sodium and potassium measurements by ion-selective electrodes in undiluted samples. Scand J Clin Lab Invest 1996; 56 (Suppl. 224): 145–60.

74. Levy GB. Determination of sodium with ion-selective electrodes. Clin Chem 1981; 27: 1435–8.
75. Lundegårdh H. The quantitative spectral analysis of the elements. Jena, Fischer 1934.
76. Maas AHJ, Siggaard-Andersen O, Weisberg HF, Zijlstra WG. Ion-selective electrodes for sodium and potassium: A new problem of what is measured and what should be reported. Clin Chem 1985; 31: 482–5.
77. Maas AHJ, Weisberg HF, Burnett RW, et al. Approved IFCC-methods. Reference method (1986) for pH measurement in blood. J Clin Chem Clin Biochem 1987; 25: 281–9.
78. Maas AHJ, Rispens P, Siggaard-Andersen O, Zijlstra WG. Calculation of true bicarbonate concentration and concentrations of other carbon dioxide species in plasma. In: Maas AHJ, Kofstad J, Siggaard-Andersen O, Kokholm G, eds. Physiology and methodology of blood gases and pH. Copenhagen, Radiometer AS: 101–8.
79. Machida Y, Nakaniski T. Utilization of bacterial xanthine oxidase for inorganic phosphorus determination. Agric Biol Chem 1982; 46: 807–9.
80. Mann CK, Yoe JH. Spectrophotometric determination of magnesium with sodium 1-azo-2-hydroxy-3-(2,4-dimethyl-carboxanilido)-naphthalene-1'-(2-hydroxy-benzene-5-sulfonate). Anal Chem 1956; 28: 202–5.
81. Markowitz M, Rotkin R, Rosen JF. Circadian rhythms of blood minerals in humans. Science 1981; 213: 672–4.
82. Massmann H. Spectrochim Acta 1968; 23B: 215–26.
83. Meier PC, Amman D, Morf WF, Simon W. In: Koryta J, ed. Medical and biological applications of electrochemical devices. New York, Wiley & Sons 1980: 18–78.
84. Morf WE. The Principles of ion-selective electrodes and of membrane transport. studies on analytical chemistry. Amsterdam, Elsevier 1981.
85. Morgan DB, Dillon S, Payne RB. The assessment of glomerular function: Creatinine clearance or plasma creatinine? Postgrad Med 1978; 154: 302–10.
86. Müller-Plathe O. A nomogramm for the interpretation of acid-base data. J Clin Chem Clin Biochem 1987; 25: 795–8.
87. Ono T, Taniguchi J, Mitsumaki H, et al. A new enzymatic assay of chloride in serum. Clin Chem 1988; 34: 552–3.

88. Payne RB. Creatinine clearance: A redundant clinical investigation. Ann Clin Biochem 1968; 23: 243–50.
89. Pei P, Cigler Ch, Vonderschmitt DJ. Evaluierung der enzymatischen Methoden zur Bestimmung von Natrium und Kalium in Plasma am Analysensystem BM/Hitachi 747. GIT Labor-Medizin 1991; 11: 482–6.
90. PetitClerc C, Solberg HE. Approved recommendation (1987) on the theory of reference values. Part 2: Selection of individuals for the production of reference values. J Clin Chem Clin Biochem 1987; 25: 639–44.
91. Röcker R, Schmidt HM, Junge B, Hoffmeister H. Orthostasebedingte Fehler bei Laboratoriumsbefunden. Med Lab 1975; 28: 267–75.
92. Schmidt LH, Wessig H, Ehrhardt W. Urinelektrolyte – Möglichkeiten und Grenzen der Interpretation. Klin Lab 1992; 38: 616–9.
93. Schmidt-Gayk H, Kasperk C. Knochen- und Mineralstoffwechsel. In: Thomas L, ed. Labor und Diagnose. 6. Aufl. Frankfurt/Main, TH-Books Verlagsgesellschaft 2005: 520–85.
94. Schultz DW, Passonneau JV, Lowry OH. An enzymic method for the measurement of inorganic phosphate. Anal Biochem 1967; 19: 300–14.
95. Schumann G, Petersen D, Büttner J. Orthophosphate determination using high performance ion chromatography and 32P liquid scintillation counting: A candidate for a reference method. Fresenius' J Anal Chem 1990; 337: 143.
96. Siggaard-Andersen O, Thode J, Fogh-Andersen N. What is "ionized calcium"? Scand J Clin Lab Invest 1983; 43 (Suppl. 165): 11–6.
97. Siggaard-Andersen O, Thode J, Fogh-Andersen N. Nomograms for calculating the concentration of ionized calcium of human blood plasma from total calcium, total protein and/or albumin, and pH. Scand J Clin Lab Invest 1983; 43 (Suppl. 165): 57–64.
98. Siggaard-Andersen O, Thode J, Wandrup J. The concentration of free calcium ions in the blood plasma, "ionized calcium". In: Siggaard-Andersen O, ed. Blood pH, carbon dioxide, oxygen, and calcium ion. Vol. 2. Copenhagen, Private Press 1981.
99. Siggard-Andersen O, Durst RA, Maas AHJ. Approved IUPAC/IFCC recommendation (1984) on physico-chemical quantities and units in clinical chemistry with special emphasis on activity and activity coefficients. J Clin Chem Clin Biochem 1984; 25: 369–91.

100. Solberg HE. Approved recommendation (1986) on the theory of reference values. Part 1: The concept of reference values. J Clin Chem Clin Biochem 1987; 25: 337–42.
101. Solberg HE. Approved recommendation (1987) on the theory of reference values. Part 5: Statistical treatment of collected reference values. Determination of reference limits. J Clin Chem Clin Biochem 1987; 25: 645–56.
102. Solberg HE, PetitClerc C. Approved recommendation (1988) on the theory of reference values. Part 3: Preparation of individuals and collection of specimens for the production of reference values. J Clin Chem Clin Biochem 1988; 26: 593–8.
103. Sonntag O. Trockenchemie. Stuttgart New York, Thieme 1988.
104. Statland BE, Bokelund H, Winkel P. Factors contributing to intraindividual variation of serum constituents: 4. Effects of posture and tourniquet application on variation of serum constituents in healthy subjects. Clin Chem 1974; 20: 1513–9.
105. Stummvoll HK. Bedrohliche Störungen des Wasser- und Elektrolythaushaltes. In: Deutsch E, Lasch HG, Lenz K, eds. Lehrbuch der internistischen Intensivtherapie. Stuttgart, Schattauer 1990.
106. Tabata M. Kido T, Totani M, Marachi T. Direct spectrophotometry of magnesium in serum after reaction with hexokinase and glucose-6-phosphatase dehydrogenase. Clin Chem 1985; 31: 703–5.
107. Thomas C, Thomas L. Niere und Harnwege. In: Thomas L, ed. Labor und Diagnose. 6. Aufl. Frankfurt/Main, TH-Books Verlagsgesellschaft 2005: 520–85.
108. Thomas L. Elektrolyt- und Wasserhaushalt. In: Thomas L, ed. Labor und Diagnose. 6. Aufl. Frankfurt/Main, TH-Books Verlagsgesellschaft 2005: 413–67.
109. Tietz NW, ed. Clinical guide to laboratory tests, 3rd edition. Philadelphia, Saunders 1995.
110. Town M-H, Jansen H, Kerscher L, Ziegenhorn J. Evaluation of enzymatic methods for the determination of sodium and potassium on the Hitachi 717 and 704. Clin Chem 1990; 36: 1069–70.
111. Truniger B, Richards P. Wasser- und Elektrolythaushalt. Stuttgart, Thieme 1985.
112. Tsunoda T, Eto T, Furukawa M, et al. Clear and colorless fluid observed during percutaneous transhepatic gallbladder drainage. Gastroenterol Jpn 1990; 25: 619–24.

113. Valentine VG, Raffin TA. The management of chylothorax. Chest 1992; 102: 586–91.
114. Velapoldi RA, Paule RC, Schaffer R, Mandel J, Moody JR. A reference method for the determination of sodium in serum. NBS Spec Public 1978: 260–60.
115. Velapoldi RA, Paule RC, Schaffer R, Mandel J, Machlan LA, Gramlich JW. A reference method for the determination of potassium in serum. NBS Spec Public 1979: 260–63.
116. Velapoldi RA, Paule RC, Schaffer R, Mandel J, Murphy TJ, Gramlich JW. A reference method for the determination of chloride in serum. NBS Spec Public 1979: 260–67.
117. Velapoldi RA, Paule RC, Schaffer R, Mandel J, Machlan LA, Garner EL, Rains TC. A reference method for the determination of lithium in serum. NBS Spec Public 1980: 260–69.
118. Vogel P, Rittersdorf W, Thym D, Bard K. Development of a potassium assay on the Reflotron. Clin Chem 1990; 36: 1070.
119. Vogt W, Oesterle B. Extreme results in electrolyte determinations. Wien Klin Wochenschr 1992; 104 (Suppl. 192): 19–24.
120. Vogt W, Braun SL. Reflections on the economic efficiency of the various methods for electrolyte determination. Wien Klin Wochenschr 1992; 104 (Suppl. 192): 29–34.
121. Walsh A. Chemical analysis by flame photometry. Spectrochim Acta 1955; 7: 108–19.
122. Walton RJ, Bijvoet OLM. Nomogram for derivation of renal threshold phosphate concentration. Lancet 1975; 2: 309–10.
123. Wandrup J, Kancir C. The concentration of free calcium ions in whole blood. Scand J Clin Lab Invest 1983; 43 (Suppl. 165): 47–8.
124. Wesinger JR, Bellorin-Font E. Magnesium and posphorus. Lancet 1998; 352: 391–6.
125. Welz B, Sperling M. Atomabsorptionsspektrometrie. Weinheim, Wiley-Vett Verlag 1999.
126. Wimberley PD, Siggaard-Andersen O, Fogh-Andersen N, Boink ABTJ. Are sodium bicarbonate and potassium bicarbonate fully dissiciated under physiological conditions? Scand J Clin Lab Invest 1985; 45: 7–10.
127. Winkel P, Statland BE, Bokelund H. The effects of time of venipuncture on variation of serum constituents. Am J Clin Pathol 1975; 64: 433–47.

Review Literature

- HG, Lenz K, eds. Lehrbuch der internistischen Intensivtherapie. Stuttgart, Schattauer 1990.
- Burtis CA, Ashwood ER, Bruns DE, eds. Tietz textbook of clinical chemistry and molecular diagnostics. 4th edition; Philadelphia, Saunders 2005.
- Fry ChH, Langley SEM. Ion-selective electrodesm for biological systems. Harwood Academic Publishers 2001.
- Heil W, Koberstein R, Zawta B. Reference ranges for adults and children – pre-analytical considerations. 8th edition. Mannheim, Roche Diagnostics GmbH 2004.
- Seldin DW, Giebisch G. The kidney. Physiology and pathophysiology. 3rd edition. Lippincott Williams & Wilkins 2000.
- Stummvoll HK. Bedrohliche Störungen des Wasser- und Elektrolythaushaltes. In: Deutsch E, Lasch HG, Lenz K, eds. Lehrbuch der internistischen Intensivtherapie. Stuttgart, Schattauer 1990.
- Thomas L, ed. Labor und Diagnose. 6. Auflage. Marburg, TH-Books Verlagsgesellschaft 2005.
- Truniger B, Richards P. Wasser- und Elektrolythaushalt. 5. Auflage. Stuttgart, Thieme 1985.
- Welz B, Sperling M. Atomabsorptionsspektrometrie. 4. Auflage. Weinheim, Wiley-VCH-Verlag 1984.

Elements

Name	Atomic Number	Symbol	International Atomic Mass*
Actinium	89	Ac	227.03
Aluminium	13	AL	26.982
Antimony (Stibium)	51	Sb	121.75
Argon	18	Ar	39.948
Arsenic	33	As	74.922
Astatine	85	At	209.99
Barium	56	Ba	137.33
Beryllium	4	Be	9.0122
Bismuth	82	Pb	207.2
Boron	5	B	10.81
Bromine	35	Br	79.904
Cadmium	48	Cd	112.41
Calcium	20	Ca	40.08
Carbon	6	C	12.011
Cesium	55	Cs	132.91
Chlorine	17	Cl	35.453
Chromium	24	Cr	51.996
Cobalt	27	Co	58.933
Copper	29	Cu	63.546
Fluorine	9	F	18.998
Francium	87	Fr	223.02
Gallium	31	Ga	69.72
Germanium	32	Ge	72.59
Gold (Aurum)	79	au	196.97
Hafnium	72	Hf	178.49
Helium	2	He	4.0026
Hydrogen	1	H	1.0079
Indium	49	In	114.82
Iodine	53	I	126.90
Iridium	77	Ir	192.22
Iron (Ferrum)	26	Fe	55.847
Krypton	36	Kr	83.80
Lanthanum	57	La	138.91
Lead (Plumbum)	82	Pb	207.2
Lithium	3	Li	6.941
Magnesium	12	Mg	24.305
Manganese	25	Mn	54.938
Mercury	80	Hg	200.59
Molybdenum	42	Mo	95.94
Neon	10	Ne	20.179
Nickel	28	Ni	58.69
Niobium	41	Nb	92.906
Nitrogen	7	N	14.007

Elements

Name	Atomic Number	Symbol	International Atomic Mass*
Osmium	76	Os	190.2
Oxygen	8	O	15.999
Palladium	46	Pd	106.42
Phosphorus	15	P	30.974
Platinum	78	Pt	195.08
Polonium	84	Po	209.98
Potassium (Kalium)	19	K	39.098
Radium	88	Ra	226.03
Radon	86	Rn	222.02
Rhenium	75	Re	186.21
Rhodium	45	Rh	102.91
Ruthenium	44	Ru	101.07
Scandium	21	Sc	44.956
Selenium	34	Se	78.96
Silicon	14	Si	28.086
Silver (Argentum)	47	Ag	107.87
Sodium (Natrium)	11	Na	22.990
Strontium	38	Sr	87.62
Sulphur	16	S	32.06
Tantalum	73	Ta	180.95
Technetium	43	Tc	98.906
Tellurium	52	Te	127.60
Thalium	81	Tl	204.38
Tin	50	Sn	181.69
Titanium	22	Ti	47.88
Tungsten	74	W	183.85
Vanadium	23	V	50.942
Xenon	54	Xe	131.29
Yttrium	39	Y	88.906
Zinc	30	Zn	65.38
Zirconium	40	Zr	91.22

*According to the SI system, mass is the preferred term. Values updated according to [50].